Beyond Interdependence

Beyond Interdependence

*The Meshing of the World's Economy
and the Earth's Ecology*

Jim MacNeill
Pieter Winsemius
Taizo Yakushiji

A TRILATERAL COMMISSION BOOK

New York   Oxford
OXFORD UNIVERSITY PRESS
1991

Oxford University Press

Oxford New York Toronto
Delhi Bombay Calcutta Madras Karachi
Petaling Jaya Singapore Hong Kong Tokyo
Nairobi Dar es Salaam Cape Town
Melbourne Auckland

and associated companies in
Berlin Ibadan

Copyright © 1991 by Jim MacNeill

Published by Oxford University Press, Inc.,
200 Madison Avenue, New York, New York 10016

Oxford is a registered trademark of Oxford University Press

Library of Congress Cataloging-in-Publication Data
MacNeill, Jim.
Beyond interdependence : the meshing of the world's economy
and the earth's ecology /
Jim MacNeill, Pieter Winsemius, and Taizo Yakushiji.
p. cm. Includes bibliographical references and index.
ISBN 0-19-507125-5;
ISBN 0-19-507126-3 (pbk)
1. Economic development—Environmental aspects
2. Environmental policy. 3. Economic history—1971–
4. International economic relations. I. Winsemius, Pieter.
II. Yakushiji, Taizō, 1944– . III. Title.
HD75.6.M33 1991 363.7—dc20 91-7348

1 3 5 7 9 8 6 4 2

Printed in the United States of America
on acid-free, recycled paper

Foreword

⎍⎍⎍⎍⎍⎍⎍⎍⎍⎍⎍⎍⎍⎍⎍⎍⎍⎍⎍⎍⎍⎍⎍

A prominent theme in reports to the Trilateral Commission, as articulated in one of our very first reports in 1974, has been that "growing interdependence and the inadequacy of present forms of cooperation are the principal features of the contemporary international order." Growing economic interdependence is what most of us have had primarily in mind in such reports and discussions. In this report, Jim MacNeill, Pieter Winsemius, and Taizo Yakushiji ask us all to go "beyond interdependence" in this too narrow economic sense, and to recognize the "meshing of the world's economy and the earth's ecology."

In part this is a physical point. As MacNeill and his coauthors vividly demonstrate, human activites have become so huge that in many instances they are of the same scale as fundamental natural processes. Critical global thresholds are being approached, and perhaps passed. And yet this is not the old argument of *Limits to Growth*—a document that the first director of the Trilateral Commission, Zbigniew Brzezinski, once termed a "pessimist manifesto." The authors instead stress the "growth of limits" through focused and urgent human efforts. Given the "growth imperative" evident in the material poverty of much of humankind, the only reasonable alternative is "sustainable development"—a concept that Jim MacNeill did so much to advance as Secretary General of the World Commission on Environment and Development (Brundtland Commission) in its landmark 1987 report *Our Common Future*.

The main theme of this Trilateral report is not the physical point, however, but rather the structural and policy point. These issues are rightly moving on to the central policy agenda. The preparation of this report to the Trilateral Commission—our first report focused so fully on these issues—is itself a reflection of this increasingly felt need for a new synthesis as we all seek to articulate central international policy needs.

MacNeill, Winsemius, and Yakushiji make clear, as have so many reports to the Trilateral Commission, the global setting in which the principal democratic industrialized countries function. One of a number of fruitful concepts developed in this report is that of the "shadow ecologies" of Trilateral countries. Economic activities centered in the European Communities, North America, and Japan cast an "ecological shadow" that is worldwide, and we need to think in such worldwide terms when we evaluate our own environmental performance. A companion point is the argument of MacNeill and his coauthors that these issues provide "growing political leverage" for developing countries:

> The active participation of developing countries is essential to the success of several . . . negotiations now underway, including those on climate change. As in the case of CFCs and ozone depletion, any reductions in fossil fuel emissions by industrialized nations could soon be wiped out by increases in a few developing nations. China alone, with one of the largest reserves of coal in the world, plans 200 new coal-fired power stations in the medium term future. With this kind of negative power, countries do not need to be rich and militarily strong to influence the behavior of great states. The problem, as experience with the Montreal Protocol demonstrates, is not that they can prevent an agreement being reached, but that they can refuse to sign, ratify or implement an agreement unless and until their economic and other concerns have been addressed.

On behalf not only of myself but also of Georges Berthoin, European chairman of the Trilateral Commission, and Isamu Yamashita, Japanese chairman, I commend this report to a wide range of readers in the Trilateral regions and beyond. While the views expressed in this report are put forward by the authors in a personal capacity and do not purport to represent those of the commission or of any organization with which the authors are associated, the chairmen of the commission hope that the report will contribute to informed discussion and treatment of the issues addressed. One particularly prominent event

on the horizon is the "Earth Summit"—the United Nations Conference on Environment and Development—which will take place in
Rio de Janiero in June 1992. MacNeill, Winsemius, and Yakushiji
look toward the Rio meeting in the concluding chapter of their remarkable
book.

David Rockefeller
North American Chairman
The Trilateral Commission

Introduction

⊓⊔⊓⊔⊓⊔⊓⊔⊓⊔⊓⊔⊓⊔⊓⊔⊓⊔⊓⊔⊓⊔⊓⊔⊓⊔⊓⊔

In June 1992 world leaders will meet in Rio de Janeiro for the largest summit conference ever held, the first truly Earth Summit. This book by Jim MacNeill, Pieter Winsemius, and Taizo Yakushiji provides a lucid exposition of the fundamental questions that prompted the United Nations General Assembly to convene the conference and of many of the difficult political and substantive issues that it will face.

The authors demonstrate that the world has now moved beyond economic interdependence to ecological interdependence—and to an intermeshing of the two. They argue persuasively that this interlocking of the world's economy and the earth's ecology "is the new reality of the century, with profound implications for the shape of our institutions of governance, national and international." They provide a fresh analysis of the implications of this new reality from a vantage point that combines breadth of vision with experience in real decision-making at the highest levels of government, industry, and international organizations. I am not surprised. I know them all and I have been privileged to work closely with the principal author, Jim MacNeill, for over two decades. He was one of my advisors when I was secretary general of the Stockholm Conference on the Human Environment in 1972. We were both members of the World Commission on Environment and Development and, as secretary general, he played a fundamental role in shaping and writing its landmark report, *Our Common Future*. He is now advising me on the road to Rio.

This book couldn't appear at a better time, with the preparations for the Earth Summit moving into high gear. No conference has ever faced the need to make such an important range of decisions, decisions that will literally determine the fate of the earth. It can build on a number of foundation stones, in particular the Stockholm Conference on the Human Environment, which put environmental issues on the global agenda, and the work of the World Commission on Environment and Development. *Our Common Future*, which is now available in over 20 languages, has sparked a global debate on sustainable development.

Twenty years after Stockholm, world leaders will meet in Rio as a direct result of the commission's recommendations. Rio will be the largest summit conference ever held, and it will have the political capacity to produce the basic changes needed in our national and international economic agendas and in our institutions of governance to ensure a secure and sustainable future for the world community. By the year 2012, these changes must be fully integrated into our economic and political life so that the world will not be forced to confront the deepening crises that will inevitably result if we fail to make the transition to sustainability.

This book extends the World Commission's analysis of the complex relationships between the environment and the economy, the changing international politics of environment (including the growing political leverage of the South), and the issues of global change. It takes account of recent events, including the Second World Climate Conference in November 1990 and the war in the Persian Gulf in early 1991.

The Earth Summit will be asked to adopt an Agenda for the 21st Century, setting out an internationally agreed work program, including targets for national and international performance on several critical issues. This "Agenda 21" cannot escape the question of reform of policies that now rig the world marketplace against both the economy and the environment. *Beyond Interdependence* provides the most compelling economic as well as environmental case for such reform that I have read. The Earth Summit will be asked to address new international conventions on climate change, forestry, and biodiversity. The authors clearly present some of the key options before the negotiators, and they discuss the danger that leaders will be tempted to adopt empty framework conventions, leaving their successors with the hard choices and the problem of finding the funds to finance sustainability measures. The options are discussed in some detail together with an analysis of their probable costs based on the latest studies. The summit will be asked to provide developing countries with access to additional financial resources to cover these costs and environmentally sound technologies to enable them to implement the conventions and to integrate environment into their future development. It will also be asked to consider far-reaching reforms of the international system.

The summit's agenda may seem to be a tall order far removed from existing political realities, particularly at a time when the attention of governments and people has been preempted, at least temporarily, by pressing crises such as the conflict in the Gulf. But this is unlikely to shake the deepening concern over the state of the environment. As the authors point out, the use of environmental destruction as a weapon in the Gulf war could serve to heighten the growing conviction that environmental risks pose the greatest threat to our common security, indeed to our very survival. The case made for broadening the concept of national security to include these threats is compelling and makes the message of the book imperative at this critical time.

The Earth Summit must succeed. There is no plausible alternative. Western leaders will have to find the political resources to demonstrate enlightened leadership. They can initiate the restructuring of international economic and political relationships needed to reverse the tragic flow of capital from the poorer to the richer countries and to ensure that developing countries get equitable access to the technologies needed to support sustainable development. Failure in these two areas, as the authors point out, would mean the failure of the summit, and that would likely cripple prospects for a new global alliance to secure the future of our planet. *Beyond Interdependence* should make a solid contribution to the success of the summit.

Maurice Strong
Secretary General
United Nations Conference on
Environment and Development

Acknowledgments

In late 1988 the Trilateral Commission asked Jim MacNeill to lead a task force on environment and development and author a report for the 1990 meeting of the commission. He agreed, and early in 1989 he consulted with Pieter Winsemius and Taizo Yakushiji, who subsequently were invited to join the task force as coauthors. While Jim MacNeill was the achitect and principal author of the report, Pieter Winsemius and Taizo Yakushiji made significant contributions throughout. Pieter Winsemius' major contribution was to chapters 4 and 5, Taizo Yakushiji's to chapter 3.

The author and coauthors owe a special debt to Edward Parson and, in appreciation, decided to recognize him as an associate author. Edward (Ted) Parson is completing his Ph.D. in public policy at the John F. Kennedy School of Government at Harvard University, where he is a teaching fellow. He began working on this project as a research assistant to Jim MacNeill but his depth of knowledge, experience, and enthusiasm enabled him to make significant contributions to all aspects of the work. Given his background, this was not surprising. He was educated at the universities of Toronto (B.Sc. in physics, 1975) and British Columbia (M.Sc. in management science, 1981), and held teaching positions in mathematics, computing, statistics, and microeconomics. He has been a consultant to the Center for Clean Air Policy in Washington, the Privy Council Office of the government of Canada, and the Office of Technology Assessment of the U.S. Congress. He has authored several papers and has written a self-study mathematics textbook. The authors are grateful for his association and contribution.

The author and coauthors are, of course, entirely responsible for the analysis and conclusions in *Beyond Interdependence*, but they were assisted by many people over many months, and wish to extend a special thanks to all of them. They held consultations with experts in London, Tokyo, Paris, and New York. In Paris Pieter Winsemius discussed aspects of the report with the following individuals: Huguette Bouchardeau, former French minister of environment; Jean Deflassieux, chairman of the Banque pour le Développement des Echanges Internationaux; François de Rose, Ambassador and former permanent representative to NATO from France; Ignacy Sachs, professor at the École des Hautes Études en Science Sociales (EHESS); and Simone Weil, member and former president of the European Parliament and former minister in the French government.

In Tokyo the authors met with the following experts: Wakako Hironaka, director of the Special Committee on Environment, Japanese House of Councillors (Komeito); Koji Kakizawa, member of the Japanese Diet and former parliamentary vice minister for environment; Takashi Kosugi, director of the Standing Committee on Environment of the Japanese Diet, as well as director, environment division, of the Policy Research Council of the Liberal Democratic party, and former parliamentary vice minister for environment; Yoshio Okawara, deputy chairman of the Trilateral Commission, executive advisor of Keidanren (Federation of Economic Organizations in Japan), and former Japanese ambassador to the United States; Saburo Okita, chairman of the Institute for Domestic and International Policy Studies, chairman of the Tokyo Conference on the Global Environment and Human Responses Toward Sustainable Development, and former minister of foreign affairs; and Masayoshi Takamura, director of the Standing Committee on Environment of the Japanese Diet.

The following individuals, whom the author and coauthors met with in New York City, reviewed and—on a personal level—provided useful insights on an early draft of the manuscript: Jesse H. Ausubel, director of studies of the Carnegie Commission on Science, Technology and Government (New York); Daniel Estey, special assistant to the administrator, U.S. Environmental Protection Agency (Washington, D.C.); Jessica Tuchman Mathews, vice president of the World Resources Institute (Washington, D.C.); Michael Oppenheimer, chief scientist of the Environmental Defense Fund (New York); Stephen H. Schneider, director of interdisciplinary climate systems at the National Center for

Atmospheric Research (Boulder, Colo.); Chris Spencer, senior advisor of the International Organizations Bureau at the Department of External Affairs (Ottawa, Canada); and Bernard Wood, director of the Canadian Institute on International Peace and Security (Ottawa, Canada).

The author and coauthors extend their appreciation to the following: Jim Bruce, chairman of the Canadian Climate Protection Board (Ottawa, Canada) and former acting deputy secretary general of the World Meteorological Organization; Charles Caccia, member of Parliament and former Canadian minister of environment and David J. R. Angell, who reviewed a later draft of the manuscript and provided a number of useful insights; Bob Munro, a consultant on environment and development policy located in Nairobi, who worked with the author on a related project concerning the 1992 Earth Summit, and Chris Spencer who was invaluable in providing bibliographic references on the rapidly proliferating series of international conferences that took place during the last few months of work on the book.

The Trilateral Commission discussed a draft of the report at its meeting in Washington, D.C., on April 22, 1990. Several members commented orally—and subsequently in writing—and the author and coauthors wish to thank them all. They owe a special debt of gratitude to Charles B. Heck, Paul Révay, and Tadashi Yamamoto, respectively the North American, European, and Japanese directors of the Trilateral Commission, as well as to Andrew V. Frankel, the assistant North American director. These people were, or seemed to be, omnipresent, providing advice on the work, organizing meetings and consultations, and offering encouragement and helpful comments on our analysis and final recommendations. Without them the work could not have been completed.

The author wishes to acknowledge the support of several people at the Institute for Research on Public Policy: Peter Dobell, vice president, persuaded him to undertake the task; and David Runnalls, director of the Program on Environment and Sustainable Development was always ready to cover for his absences. Thanks also go to Jeffrey Holmes, director of communications, for his systematic editing of the final text.

Lastly, the author owes a debt of thanks to his wife, Phyllis MacNeill, for her unfailing patience in reading and providing helpful comments on repeated drafts of the text as they rolled off his computer.

Contents

⊓⊔⊓⊔⊓⊔⊓⊔⊓⊔⊓⊔⊓⊔⊓⊔⊓⊔⊓⊔⊓⊔⊓⊔⊓⊔⊓⊔

Abbreviations

ꓕꓥꓥꓥꓥꓥꓥꓥꓥꓥꓥꓥꓥꓥꓥꓥꓥꓥꓥꓥꓥꓥ

ASEAN—Association of South East Asian Nations
CARICOM—Caribbean Community
CEC—Commission of the European Communities
EC—European Communities
ECE—Economic Commission for Europe (U.N. organization)
EFTA—European Free Trade Association
EPA—Environmental Protection Agency
FAO—Food and Agricultural Organization
GATT—General Agreement on Tariffs and Trade
G7—Group of Seven
IADB—Inter-American Development Bank
IEA—International Energy Agency
IIASA—International Institute for Applied Systems Analysis
IMF—International Monetary Fund
IPCC—Intergovernmental Panel on Climate Change
ITTO—International Tropical Timber Organization
IUCN—International Union for the Conservation of Nature
NPT—Non-Proliferation Treaty
ODA—Official Development Assistance
OECD—Organization for Economic Cooperation and Development
PACD—Plan of Action to Combat Desertification
PECC—Pacific Economic Cooperation Conference

PPP—Polluter Pays Principle

SAARC—South Asian Association for Regional Cooperation

SADCC—Southern Africa Development Coordination Conference

TFAP—Tropical Forests Action Plan

UNCTAD—United Nations Conference on Trade and Development

UNDP—United Nations Development Program

UNEP—United Nations Environment Program

UNESCO—United Nations Educational, Scientific and Cultural
Organization

USAID—United States Agency for International Development

WCED—World Commission on Environment and Development

WEDF—World Environment and Development Forum

WHO—World Health Organization

WMO—World Meteorological Organization

WRI—World Resources Institute

Beyond Interdependence

1

The Growth Imperative and Sustainable Development

ᴸᴸᴸᴸᴸᴸᴸᴸᴸᴸᴸᴸᴸᴸᴸᴸᴸᴸᴸᴸᴸᴸᴸᴸᴸᴸ

The last decade of the twentieth century is a time of great promise, great risk, and great complexity. Events are accelerating on several fronts simultaneously—economic, ecological, and political—and are forcing profound changes in the relationships among peoples, nations, and governments.

The gains in human welfare over the past few decades have been breathtaking and, if we continue to avoid world-scale conflict, the potential for future gains is even more awesome. Many new and emerging technologies in biology, materials, construction, satellite monitoring, and other fields offer great promise for increasing the production of food, developing more benign forms of energy, raising industrial productivity, conserving the earth's basic stocks of natural capital, and managing the environment. With global communications and ever-greater access to information, people can now begin to exercise responsibility for every part of the planet.

Past gains, however, have been accompanied by vast increases in the the scale of human impact on the earth. Since 1900, the world's population has multiplied more than three times. Its economy has grown twentyfold. The consumption of fossil fuels has grown by a factor of 30, and industrial production by a factor of 50. Most of that growth, about four-fifths of it, occurred since 1950. Much of it is unsustainable. Earth's basic life-supporting capital of forests, species, and soils is being depleted and its fresh waters and oceans are being degraded at an accelerating rate. Even the ozone shield that protects all life from the

3

sun's more deadly rays is being slowly consumed. And now the earth is threatened by a rapid rise in global temperatures and sea levels—greater, perhaps, in the next century than in the 10,000 years since the last ice age.

Since World War II, governments have been preoccupied with economic interdependence, the coupling of local and national economies into a global system. But the world has now moved beyond economic interdependence to ecological interdependence—and even beyond that to an intermeshing of the two. The earth's signals are unmistakable. Global warming is a form of feedback from the earth's ecological system to the world's economic system. So is the ozone hole, acid rain in Europe, soil degradation in Africa and Australia, deforestation and species loss in the Amazon. To ignore one system today is to jeopardize the others. The world's economy and earth's ecology are now interlocked—"until death do them part," to quote one of Canada's industrial leaders.[1] This is the new reality of the century, with profound implications for the shape of our institutions of governance, national and international. It raises fundamental questions about how economic and political decisions are made, and their implications for sustainability.

The tidal shift in East–West relations should make it easier to address this new reality. For more than 40 years, world affairs have been dominated by the contest between East and West. The goal of each was to contain the expansion of the other, and that goal was pursued at enormous cost in human and financial resources and in opportunities foregone. The end of the cold war raised the possibility that money once used for military purposes could be channeled into constructive cooperation between the superpowers and their allies on international development, the protection of the environment, and the building of a durable peace. Instead, the 1990s commenced with growing concern about political and economic instability in Central and Eastern Europe (including the Soviet Union) and a major conflict in the Persian Gulf. These events have postponed at least temporarily the possibility of significant new resources for the critical issues of development and environment. Yet, public concern about the environment is intense and continued ecological decline in the Soviet Union and Eastern Europe and the use of environmental destruction as a weapon in the Gulf could intensify the pressure for action.

The issues of development and environment are beginning to reshape national and international affairs and they could well become even more critical during the next two decades. If human numbers do double again, a five- to tenfold increase in economic activity would be required to enable them to meet their basic needs and minimal aspirations. Aspirations are just as important as needs. A five- to tenfold increase translates into a colossal new burden on the ecosphere and raises the question of sustainability. Is there, in fact, any way to multiply economic activity a further five to ten times, without it undermining itself and compromising the future completely? Can growth on these orders of magnitude be managed on a basis that is sustainable? The question of sustainability has been forced front and center by the acceleration of events. It will preoccupy governments, industry, and our institutions of higher learning well into the next century.

In June 1992 world leaders will assemble in Rio de Janeiro at the U.N. Conference on Environment and Development for an Earth Summit. Their objective will be to launch a gradual shift to more sustainable forms of development. What issues will confront them? Are there solutions to these issues? Can future development be made more sustainable? What is the cost of action—and inaction? Can we marshall the political resources to act? These and many other questions are addressed in the pages that follow.

The Growth Imperative

The world is on history's fastest growth track. As noted, population will very likely double within the next half century. Governments could act to stabilize population at lower levels, but with one-third of the world's people under 15 years of age, even the most vigorous policies will not avert rapid population growth and the accompanying need for large increases in production.

How large an increase? If current forms of development were employed, a further five- to tenfold boost in economic activity would be required over the next 50 years to meet the needs and aspirations of 10 billion people.[2] Such an increase may appear colossal, but it reflects the continuance of annual growth rates of 3.2 to 4.7 percent. While difficult to maintain, political imperatives make it impossible for most governments to aspire to less. And few opposition parties can promise less if the government fails. In fact, in many developing countries these growth rates are hardly enough to keep up with projected rates of population growth, let alone to reduce poverty levels.

Many governments today fail to enable their people to meet even their most basic needs. Over 1.3 billion lack access to safe drinking water; 880 million adults cannot read or write; 770 million have insufficient food for an active working life; and 800 million live in "absolute poverty," lacking even the most rudimentary necessities. Each year 14 million children—about 10 percent of the number of children born annually—die of hunger.[3]

In part this reflects the fact that during the past 40 years economic growth has been concentrated in the North. With 25—soon to be 20— percent of the world's population, industrialized countries consume about 80 percent of the world's goods.[4] That leaves more than three-quarters of the world's population with less than one-quarter of its product. And the imbalance is getting worse, increasing tensions with the South and leaving increasing numbers of people poor and vulnerable.

To reduce rather than increase mass poverty during the first part of the next century, developing countries would have to increase per capita income by a minimum of 3 percent per year (see Appendix 1). In addition, they would need to pursue vigorous policies to reduce presently wide and growing income disparities. Per capita income growth of 3 percent a year would require overall national income growth of around 5 percent a year in the developing countries of Asia, 5.5 percent in Latin America, and 6 percent in Africa and West Asia.

During the 1960s and 1970s, many countries in these regions experienced growth of this magnitude, and a few still do. During the 1980s, however, the situation in many developing countries changed dramatically for the worse, not just economically, but also ecologically. Population growth outstripped economic growth and two-thirds of these countries suffered falls in per capita income, some as great as 25 percent. Deteriorating terms of trade, including unstable commodity prices, growing protectionism in developed market economies, and stagnating flows of aid all combined to force attention on short-term crises rather than long-term development.

Crossing Critical Thresholds

In face of the finiteness of the earth's ecosystems, we must ask whether economic growth of the magnitude required over the next half century

TABLE 1 Resource Dependence of Selected Developing Countries

Economies	Agricultural Production as a Percentage of GDP		Employment in Agriculture as a Percentage of Total Employment		Exports of Primary Products as a Percentage of Total Exports	
	1965	1986	1965	1980	1965	1986
LOW INCOME						
Burma	35	48	64	53	99	87
China	39	31	81	74	54	36
India	47	32	73	70	51	38
Indonesia	56	26	71	57	96	79
Sri Lanka	28	26	56	53	99	59
Ethiopia	58	48	86	80	99	99
Ghana	44	45	61	56	98	98
Kenya	35	30	86	81	94	84
Nigeria	53	41	72	68	97	98
Tanzania	46	59	92	86	87	83
MIDDLE INCOME						
Bolivia	23	24	54	46	95	98
Colombia	30	20	45	34	96	82
Costa Rica	24	21	47	31	84	65
Thailand	35	17	82	71	95	58
Senegal	25	22	83	81	97	71
Zimbabwe	18	11	79	73	71	64
INDUSTRIAL MARKET						
Canada	6	3	10	5	63	36
Japan	9	3	26	11	9	2
Spain	15	6	34	17	60	28
United States	3	2	5	4	35	24

Source: Compiled from World Bank, World Development Report, 1988, Oxford University Press, 1988.

The agricultural sector is comprised of agriculture, forestry, fishing and hunting. Primary products, in addition to agriculture, include fuels, minerals, and metals.

is possible at all. As can be seen from Table 1, the economies of most Third World countries, and parts of many industrialized countries, are based on their natural resources. Their soils, forests, fisheries, species, and waters make up their principal stocks of economic capital. The overexploitation and depletion of these stocks can provide them with financial gains in the very short term, but will result in a steady reduction of their economic potential over the medium and longer term.

Even at present levels of economic activity, there is growing evidence that certain critical global thresholds are being approached, perhaps passed. The evidence comes first from comparing the total scale of material and energy flows from human activities with the scale of corresponding flows in natural systems; and second from cataloging the nature and magnitude of environmental changes that are attributable to human activity. Both people and natural systems transport and transform vast quantities of carbon, nitrogen, and other materials, as well as energy. People do it through large-scale economic activities such as agriculture, mining, energy use, and waste disposal. Natural systems do it by transforming the sun's energy into plant growth and decay, and by cycling materials through atmospheric and ocean currents.

Throughout history, the scale at which people transformed energy and materials has been minuscule in comparison with nature. Recently, however, this has changed and the relationship between human society and the planet has been transformed profoundly. Human activities have become so huge that in many instances they are of the same scale as fundamental natural processes. Plant products, nitrogen, and carbon provide the most striking examples. Humans now take or transform for their use almost half of the plant material fixed by photosynthesis over the earth's entire surface.[5] When it comes to nitrogen, humans today fix almost as much nitrogen in the environment as does nature, largely because of artificial fertilizers.[6] We are also altering the carbon cycle, having increased the carbon dioxide content of the atmosphere by 25 percent in the past 130 years. Currently, we are putting carbon into the atmosphere at the rate of about 7 to 8 billion tons per year, about 7 percent of the total natural carbon exchange between the atmosphere and the oceans.[7]

When human and natural processes approach similar magnitudes, human-induced disruption of global systems becomes a serious probability. We would expect to observe increasingly complex, possibly irreversible syndromes of environmental degradation, and, indeed, we are now observing just that. Pollution problems that were once local are now regional and even global in scale (e.g., acid rain, involving the entire continents of North America and Europe). Environmental effects that once appeared simple and trivial are now seen to be complex and substantial, slowly changing ecological systems critical to economic development and to life itself (e.g., the dispersion of chemicals in soils and water and their concentration in the food chain). Acute episodes of reversible damage which were once thought to affect mainly the current generation are now seen to affect seriously the health and welfare of generations yet unborn (e.g., chemicals in body tissue,radiation exposure, the loss of genetic resources). Cities and settlements that once grew in response to employment generation and in pace with basic services are now exploding in developing countries, overwhelming both jobs and services.[8]

The loss of forests provides a stark example of the scale and rapidity of human transformation of the earth's land surface. Since 1850 the earth's forest cover has been reduced from 6 billion to 4 billion hectares. During the past four decades, the rate of deforestation has accelerated sharply, especially in developing (mostly tropical) countries, where about 60 percent of the remaining forests are found. Forty years ago, Ethiopia had a 30 percent forest cover; twelve years ago, that had fallen to 4 percent; today, it may be 1 percent. Until this century, India's forests covered more than half the country. Today, they are down to 14 percent. In 1961 Thailand's forests covered 53 percent of the country. Fifteen years later (1986) they were down to 29 percent—and going fast. In 1988 tropical forests were disappearing at a rate estimated conservatively at over 20 million hectares per year, about the size of the United Kingdom. Brazil alone may be losing over 8 million hectares annually.[9]

Many of the 24 advanced industrialized nations of Western Europe, North America, and Austral-Asia (Japan, Australia, and New Zealand) that belong to the Organization for Economic Cooperation and Development (OECD) are experiencing net forest loss through a combination of overcutting, bad forest management, and damage from acid rain. So are all of the nations of Eastern Europe. Forest loss due to acid rain damage is particularly severe in Europe. More than 50 million hectares (35 percent) of Europe's forests are damaged, dead, or dying. In 1988 the U.N. Economic Commission for Europe (ECE) reported that in 22 areas (mostly nations), 30 percent or more of the forests were damaged, and that in eight areas (again mostly nations), 50 percent were damaged.[10] The evidence is not all in, but many reports show soils in parts of Europe becoming acid throughout the tree-rooting layers.[11] Large parts of Western and Eastern Europe may be experiencing irreversible acidification, the remedial costs of which could be beyond economic reach.[12]

The destructive effects of forest loss on both the ecology and economy are becoming increasingly evident. The removal of upland forests is linked directly to increased lowland flooding and silting of rivers and reservoirs. This is a growing problem in both developed and developing countries, but it is especially acute in areas such as the lowland plains of Pakistan, India, and Bangladesh, where the welfare of more than 400 million lowland people depends on how 46 million hill dwellers manage their land. In Europe and North America, too, deforestation can trigger a chain reaction leading to increased erosion, silting of rivers, and flooding of valley farmlands and towns. Many believe that the most serious consequence of disappearing forests and other wildlands is the loss of habitat for species and the accelerating and irreversible destruction of genetic diversity. Estimates of the total number of species on earth range from 5 to 30 million, with estimated losses by 2000 of 15 to 30 percent of the total. This mass extinction robs present and future generations of genetic material from which to provide medicines and with which to improve crop varieties to make them more resistant to pests, disease, and climate change. Deforestation also results in a net release of carbon dioxide from vegetation and soil, as these carbon reservoirs undergo burning and decomposition.

The loss of soils and growing shortages of water in the face of escalating world food needs provide an equally stark picture. An area larger than the African continent and inhabited by more than 1 billion people is now at risk of turning into desert. Recent studies estimate that 6 to 7 million hectares of agricultural land are lost annually to erosion, and about 1.5 million more, mostly irrigated land, is lost to waterlogging, salinization, and alkalization. Every year the world's population increases by 90 million; and every year 25 billion tons of topsoil are lost, roughly the amount that covers Australia's wheatlands.[13] Water use has doubled at least twice in this century and could double again over the next two decades. Yet, in 80 developing countries, with 40 percent of the world's population, water is already a serious constraint on development. Land degradation and water shortages have created millions of environmental refugees worldwide, leading to large destabilizing flows across borders and into already overcrowded Third World cities. In 1984–1985 there were 10 million such refugees, two-thirds of all refugees worldwide.[14]

Perhaps the most sobering indications of human-induced environmental stress are the changes occurring in the atmosphere. Chlorofluorocarbons (CFCs) and other gases released by humans are consuming the protective ozone layer, which serves to prevent the sun's harmful ultraviolet rays from reaching the surface of the earth. CFCs are also one of the more significant "greenhouse gases," accumulations of which are increasing the heat-trapping ability of the atmosphere, threatening major disruptions of the world's climate.

In 1988 scientists reported that the average global concentrations of ozone in the stratosphere had fallen by approximately 2 percent between 1969 and 1986. The decline varied by latitude and season, but the most heavily populated regions of North America, Europe, and the Orient experienced a year-round decline of 3 percent and a winter decline of 4.7 percent.[15] The consequences are life threatening: exposure to increased ultraviolet radiation not only promotes skin cancer, but also reduces crop yields. It also threatens oceanic life-support systems because of the sensitivity of phytoplankton to ultraviolet rays.

Greenhouse gases come mostly from natural sources but, as noted, man-made sources have become significant. About half of the human-induced increase comes from carbon dioxide (CO_2), principally from fossil-fuel burning (coal, oil, natural gas) and deforestation (see Table 2). CFCs used in foams, aerosols, refrigerants, and solvents are responsible for about one-quarter of the human-induced greenhouse effect. Methane (CH_4) from wetlands, rice paddies, livestock, and fossil fuels is also a significant contributor, as is nitrous oxide (N_2O) from fertilizers, deforestation, and—again—fossil fuels. Other trace gases include carbon monoxide (CO) and ozone (O_3).

During the 160,000 years prior to the Industrial Revolution, the amount of carbon dioxide in the atmosphere did not exceed 280 parts per million. For thousands of years prior to the mid-1800s, it remained around 275 ppm. Since then, however, it has increased to 350 ppm and it is presently increasing at 1.8 ppm or 0.5 percent per year. Concentrations of the other greenhouse gases are also increasing: methane at 1 percent per year, CFCs at 4 percent, and nitrous oxide at 0.25 percent per year.

According to present scientific knowledge, the increase in greenhouse gases since the 1850s should have caused a 0.3 to 0.7 degree Celsius rise in average global temperature to date and a commitment to a 0.7 to 1.5 degree centigrade rise in the future. This distinction is important, because the warming of the earth is a slow process, mainly due to the effect of the oceans. In its recent scientific assessment, the Intergovernmental Panel on Climate Change (IPCC) projected increases in all greenhouse gas emissions over the next century assuming various scen-

TABLE 2 Major Greenhouse Gases Affected by Human Activities

	Carbon Dioxide (ppmv)	Methane (ppmv)	CFC-11 (pptv)	CFC-12 (pptv)	Nitrous Oxide (ppbv)
Pre-Industrial (1750–1800) Atmospheric Concentration	280	0.8	0	0	288
Present Day (1990) Atmospheric Concentration	353	1.72	280	484	310
Current Rate of Change/Year	1.8 (0.5%)	0.015 (0.9%)	9.5 (4%)	17 (4%)	0.8 (0.25%)
Atmospheric Lifetime (Years)	(50–200)*	10	65	130	150
Current Greenhouse Contribution (percent)	55	15	17 (CFCs 11 & 12) 7 (other CFCs)		6
Principal Sources	Fossil fuels (coal, oil, natural gas) deforestation	Wetlands, rice paddies fossil fuels, livestock	Foams, aerosols, refrigerants, solvents		Fossil fuels, fertilizers, deforestation

Sources: Intergovernmental Panel on Climate Change, *Climate Change: The Scientific Assessment*, Report Prepared for IPCC by Working Group I, Editors, Houghton, Jenkins and Ephraums, Cambridge University Press, 1990.

*No single value can be given because of complex way CO_2 is absorbed by the biosphere and oceans.

Ppmv = parts per million by volume; ppbv = parts per billion (thousand million) by volume; pptv = parts per trillion (million million) by volume.

arios of economic activity. If no steps are taken to limit greenhouse gas emissions, that is, if we continue with "business as usual," global mean temperatures will increase between 2.6 and 5.8 degrees Celsius over the next century, and sea levels will rise between 30 and 100 centimeters, or up to one meter.[16]

This does not reflect the full range of possible global warming and sea level rise in the twenty-first century. In 1987, a workshop of experts estimated the uncertainty associated with rates of temperature change and sea-level rise. Their estimates, which are based on slightly different scenarios than those employed by the IPCC, are shown in Figures 1.1a and b.[17] In each case, the middle scenario reflects a continuation of present emission trends except for CFCs, where the experts assumed the implementation of the 1987 Montreal Protocol of the Vienna Convention on the Protection of the Ozone Layer.[18] The upper scenario represents the change that could result from a radical expansion in the use of fossil fuels and other activities emitting greenhouse gases. The lower scenario represents the changes that could result from a strong global effort to reduce greenhouse gas emissions. As regards rates of temperature change, estimates range from a low rate of 0.06 degrees Celsuis per decade to a high of 0.8 degrees per decade, with a middle scenario of 0.3 degrees. As regards rates of sea-level rise, the corresponding estimates range from a drop of 1 centimeter per decade to a rise of 24 centimeters per decade, with a middle scenario of 5.5 centimeters per decade.[19]

These wide ranges reflect the uncertainties associated with future patterns of fossil fuel use, rates of deforestation, and other activities leading to greenhouse gas emissions. They also depict the response of climate systems to a given level of greenhouse gas emissions, but do not include uncertainty from all of the possible feedbacks in the climate system as a result of global warming.[20] In the judgment of the experts, there is a 50 percent chance that the actual path of climate change could lie below the middle scenario and a 90 percent chance that it will fall between the upper and lower curves. The possibility remains, however, that unforeseen circumstances could lead to climate change outside the curves depicted in the figures.

Even a few tens of centimeters of sea-level rise can have serious consequences, particularly when coupled with the possibility of increased storm surge (see Figure 1.2). The nations of the Caribbean, for example, depend heavily on their beaches for employment and foreign

FIGURE 1.1 Scenarios of temperature and sea-level rise: (a) Global temperature change (°C); (b) Global sea-level change (cm)

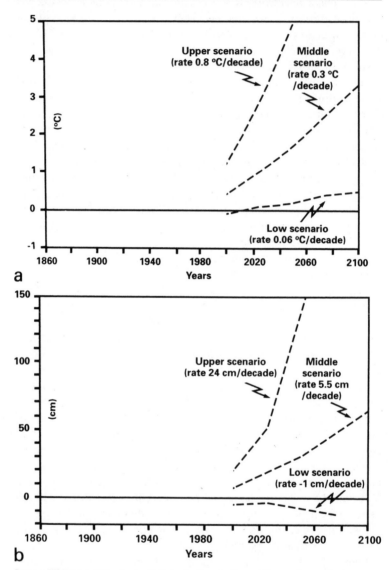

Source: Jill Jaeger, "Developing Policies for Responding to Climate Change," a summary of the discussions and recommendations of the workshop held in Villach on September 28–October 2, 1978, and Bellagio on November 9–13, 1987, under the auspices of the Beijer Institute, Stockholm. World Meteorological Organization and United Nations Environment Programme, April 1988, pp. 4–5. Courtesy of World Meteorological Organization.

FIGURE 1.2 Coastal areas vulnerable to sea-level rise

Source: Delft Hydraulics, The Netherlands.

exchange. Their beaches average only 20 meters in width, and will withdraw by one meter for every centimeter rise in sea levels. A rise of up to 1 meter by the end of the century, near the upper end of the IPCC's projection, would be catastrophic for many nations. Many small island nations could virtually disappear. Prime Minister Bob Hawke of Australia has offered the residents of several threatened South Pacific island states the opportunity to be resettled in Australia. Other countries could lose coastal areas that support much of their population and economic infrastructure. The projected increase in temperatures and sea levels, of course, does not stop at the end of the century. It will continue unless governments can marshal the political resources required to institute strong limitation measures.

A rise of 1.5 to 2.0 meters would flood 20 percent of the land in Bangledesh, inundate most of the islands of the Maldives, and threaten nearly half of the population and economic production of Thailand. A Commonwealth prime ministers' task force report, completed in 1989, speaks of "far-reaching social and economic effects on low-lying areas, as in Guyana . . . Kiribati, Tuvalu, and other Commonwealth countries. . . . They face the prospect not only of flooding from the sea and greater risk of storm surges, but deeper flooding on inland plains. . . . There are potential implications . . . for agriculture; fresh water supply threatened with salinization; the siting of towns, factories, power plants and airports; and hazardous waste disposal."[21]

Global warming will also increase the probability of severe droughts in the agricultural heartlands of Australia, Europe, North America, and the Soviet Union, disrupting areas that provide a large proportion of the world's cereal harvests. Changes in the hydrological cycle may render billions of dollars of water management and hydroelectric projects prematurely obsolete. Climate change brings into question the economic viability of hundreds of billions of dollars worth of investment now planned on the implicit assumption of stable climate patterns. Changes in both temperature and precipitation patterns could totally undermine elaborate plans for the "sustained yield" management of forests, since climate changes may advance faster than some forest species can migrate. Overall, climate change could provoke "potentially severe economic and social dislocation . . . which will worsen international tensions and increase the risk of conflicts among and within nations. These . . . changes may well become the major non-military threat to international security and the future of the global economy."[22]

As indicated by figures 1.1a and b, there are substantial uncertainties about the projected rates of global warming and sea-level rise. Disagreement exists as to the adequacy of the data base, the significance of the models' omission of realistic ocean-atmosphere coupling, the time lags between changes in atmospheric concentration of greenhouse gases and changes in climate, the feedback from clouds (which could either increase or decrease the rate of warming), and whether or not the temperature record of the past 100 years already shows greenhouse warming. But the basic science is not in dispute. The heat-trapping properties of the greenhouse gases are well known, and their buildup in the atmosphere is well documented.

There are good reasons to take atmospheric models very seriously. They have been tested on their ability to predict large changes in climate, and they have performed well. They predict accurately the seasonal variations in climate between the Northern and Southern hemispheres, and are even more accurate at predicting the radically different climates of Venus and Mars. More significantly, perhaps, the record of the earth's climatic history that can be reconstructed from the evidence of fossils and glaciers shows a nearly perfect correlation between carbon dioxide concentration and temperature. Figure 1.3 shows the CO_2 and temperature record constructed from air bubbles trapped in an Antarctic ice core.

Uncertainty is a two-edged sword. The presence of uncertainty in the 2.6 to 5.8 degree range, and even in the wider range shown in figures 1.1a and b, means that climate change may be less and come later than we think, as advocates of inaction assert; but it also means that it may be more and come sooner, as was the case with models predicting rates of stratospheric ozone depletion. Uncertain science cannot be taken as a reason for inaction. It is irresponsible to focus on ''worst-case'' results in order to justify major policy changes, but it is equally irresponsible to focus on ''best-case'' results in order to justify a complacent ''wait-and-see'' posture.

We take huge risks if we continue to increase concentrations of greenhouse gases at the present rate and, in the face of uncertainty, the prudent course is to take some action early in the hope of cutting off the worst possible outcomes. The imprudent course is to do nothing, awaiting a complete confirmation of the models. As a recent World Bank paper states: ''When confronted with risks which could be menacing, cumulative and irreversible, uncertainty argues strongly in favour

FIGURE 1.3 Carbon dioxide and temperature change.

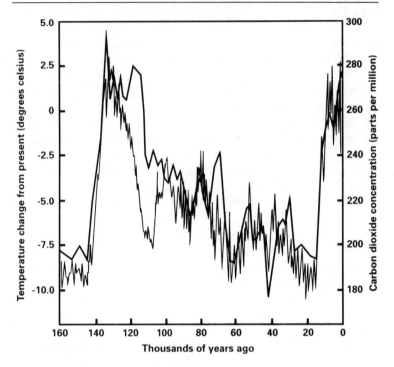

Source: Stephen H. Schneider, "The Changing Climate," *Scientific American*, September 1989. Courtesy of Stephen H. Schneider.

Carbon dioxide and temperature have been very closely correlated over the past 160,000 years. This record is based on evidence from Antarctica and shows how the local temperature (the more jagged line) and atmospheric carbon dioxide concentrations rose nearly in step as the ice age receded, beginning about 130,000 years ago, fell almost in synchrony at the onset of the new glacial period, and rose again as the ice age retreated about 10,000 years ago.

of prudent action and against complacency."[23] The IPCC estimates that a reduction of over 60 percent in carbon dioxide emissions would be required to stabilize atmospheric concentrations at 1990 levels. Even greater reductions of CFCs (70–85 percent) and nitrous oxides (70–80 percent) would be necessary, along with a smaller (15–20 percent) reduction in methane.[24] The measures needed to achieve a reduction in fossil fuel emissions of carbon dioxide and nitrous oxides are well known and could be justified largely by the reductions in air pollution and acid rain that would follow, as well as the increases in macroeconomic efficiency. The industrialized nations have already agreed to phase out most CFCs by 2000 but, as yet, we do not know how to reduce methane. To be effective, the measures to reduce CFCs and CO_2 must be initiated now, and research on methane must be stepped up. Both the physical and political systems involved are huge and complex. There will be long lags both between a change in political rhetoric and any real change in policies, and between policy change and any real effect on rates of climate change.

Conditions for More Sustainable Forms of Development

Contrasting the scale of the world's developmental needs with the evidence of the earth's environmental limits presents governments with a paradox. The growth required over the next few decades to meet human needs and aspirations translates into a colossal new burden on the ecosphere. If, for example, nations were to employ current forms of energy development, energy supply would have to increase by a factor of 5 for developing countries, with their present populations, to experience the level of consumption now prevailing in the industrialized world.[25] Similar factors can be cited for food, water, shelter, and the other necessities of life. Critical life-support systems would probably collapse long before those levels were reached.

Can growth on the scale needed to meet future needs and aspirations be managed on a basis that is sustainable? This question, the question of sustainability, has been forced front and center by the acceleration of events. The answers to it are not evident. The horizon may glow with technological opportunities, but the obstacles to sustainability are not mainly technical; they are social, institutional and political. Given the constraints on social, institutional and political change, no one can rule out a future of progressive ecological collapse. The Four Horsemen of the Apocalypse—war, famine, pestilence, and death—are galloping through parts of Africa, Asia, and Latin America, and they will surely remain active, spurred on by increasing poverty and greed, and by policy and institutional failures. Threats to the peace and security of nations from environmental breakdown are increasing at a frightening pace.

Conflicts based on climate change, environmental disruption, and water and other resource scarcities could well become endemic in the world of the future.

Overcoming the obstacles to sustainable development will require political vision and courage in policy and institutional change on a scale not seen in this century since the aftermath of World War II. The concept of "sustainable development"—not the type of development that dominates today, but development based on forms and processes that do not undermine the integrity of the environment on which they depend—may provide the needed guide for this vision, as well as a pragmatic approach to the paradox of growth and limits.

Adding the condition of sustainability to the goal of development may seem to state the obvious. Indeed, theories of economic development have always emphasized the maintenance and accumulation of capital through investment, which is implicitly a narrow sustainability condition. But the projected pace of global change indicates that a broader set of conditions must be met if development is to be sustainable. The World Commission on Environment and Development (WCED) put forward a number of such conditions, which they referred to as "strategic imperatives for sustainable development." These included growth sufficient to meet human needs and aspirations and a more equitable distribution of the proceeds of growth within and between nations. Our declining stocks of ecological capital must be conserved and increased, and the amount of energy and natural resources contained in every product must be reduced. Most of all, environment and economics must be integrated in all of our major institutions of decision-making—government, industry, and the home (see Appendix 2).

In addition, the WCED stressed that sustainable development depends on a political system that ensures effective citizen participation in decision-making (in other words, human rights and democracy), an economic system able to generate surpluses on a sustainable basis, and an administrative system that is flexible, with a built-in capacity for self-correction. The productivity of ecosystems and environmental dynamics must also be maintained. In addition to changes in policies and processes of decision-making, this will require new and better technologies in most fields. Major global thresholds must be avoided by comfortable margins, to pay due respect to our ignorance and to ensure against external shocks. And these conditions must be satisfied over larger spatial scales and longer time scales than our management institutions have yet developed the ability to deal with. A greatly improved scientific capacity to anticipate and manage risk is imperative.[26]

At the moment, governments are moving in the wrong direction on a more equitable distribution of the proceeds of growth between and within nations. As Table 3 illustrates, the cumulative debt of developing countries surpassed $1 trillion in 1986, and interest payments now exceed $70 billion a year. Net transfers of capital to developing countries turned negative in 1982. The poorer nations transferred over $50 billion to the richer nations in 1989.[27] And that is only what the World Bank counts. In addition, today's trade patterns contain a massive transfer of the environmental costs of world GNP to the resource-based economies of developing countries. These costs were estimated at $14 billion a year in 1980 (counting only pollution but not counting resource depletion), which was around one-half of the development assistance then flowing in the other direction.[28] The situation has grown much worse in the past decade. Yet, we have just witnessed an entire round of negotiations to reform the General Agreement on Tariffs and Trade (GATT), the so-called Uraguay Round, without any serious discussion of the environment.

Governments are also moving backward on population. A sustainable planet depends on a significant and rapid reduction in high rates of population growth, in rich and poor countries alike. The issue is not simply the number of people, for a child born in a rich, industrialized

TABLE 3 The Deteriorating Financial Position of Developing Contries, Developing Country Debt Stocks, and Associated Financial Flows, 1980–89 (U.S.$ billions)

	1980	1981	1982	1984	1986	1988	1989*
Total debt stocks (year end)	572	753	819	855	1047	1156	1165
Total debt flows							
1. Disbursements	112	124	108	97	103	108	111
2. Principal repayments	46	49	46	50	76	88	86
3. Net flows (1 − 2 = 3)	66	75	62	47	27	20	26
4. Interest payments	47	68	65	69	65	72	77
5. Net transfers (1 − 2 − 4 = 5)	+19	+6	−2	−22	−38	−52	−52

Source: World Bank, "World Debt Tables, 1989–90, External Debt of Developing Countries," Vol. 1: Analysis and Summary Tables, p. 78, the World Bank 1989, Washington, D.C.

*Projected

country places a much greater burden on the planet than one born in a poor country. The industrialized world has found that development coupled with access to family planning is the best means of population control. Even negative rates of population growth have been achieved where development and family planning were accompanied by urbanization, rising levels of income, improved education, and the empowerment of women. During the 1990s, however, 3 billion young people—about equal to the world's population in 1960—will enter their childbearing years, most of them without access to dependable means of birth control.[29] In spite of this, some Western nations are reducing rather than increasing their contributions to population programs and restricting rather than facilitating the distribution of new family planning technologies. Some are even supporting fertility programs! Unless Western countries, and in particular the United States, provide massively increased financial and political support for population programs, it is unlikely that currently unsustainable rates of population growth can be reversed.

A critical condition for sustainable development is the maintenance of a community's or a nation's basic stock of natural capital. Is this a reasonable condition? Can the world's expanding economies begin to live off the interest of the earth's stock of renewable resources without encroaching on its capital? There is no doubt that, at the moment, nations are consuming their natural capital at an accelerating pace, but the question is open. If use of the earth's basic economic capital is to be brought within the capacity of natural systems to generate it, governments will need to increase by several orders of magnitude their support for strategies aimed at abating pollution, protecting and preserving critical stocks of natural capital, and restoring and rehabilitating those assets that have already been depleted.

What does this condition mean for nonrenewable resources, where extraction and use must reduce the capital available? It is often assumed that the supply will ultimately limit the use of nonrenewable resources. This may be the case for some nonrenewable resources, but for others, such as coal, it seems more probable that their use will be limited by their negative impacts on renewable resources—soils, waters, forests, and the atmosphere. Increasingly, therefore, the extraction, use, and disposal of nonrenewable resources will have to take into account their impacts on renewable resources. The rates of depletion of nonrenewable resources should also take into account the criticality of the resource, the availability of technologies for minimizing depletion, and the likelihood of substitutes being available.

If nations are to stop depleting their basic stocks of ecological capital, governments will need to reform those public policies that now actively encourage the infamous *des: de*forestation, *de*sertification, *de*struction of habitat and species, *de*cline of air and water quality. Virtually all governments today pay lip service to the market. And then they intervene to distort it in ways they find politically convenient. Subsidies, tax abatements, fiscal incentives, price supports, tariffs, and trade quotas of all kinds can distort prices and trading patterns in ways that are economically perverse and encourage unsustainable forms of development. They often rig the market not only against the economy, but also against the environment and, ultimately, against development itself.

Take deforestation. Forests in developing countries are being rapidly converted due to tax concessions, "sweetheart leases," and direct subsidies that encourage increased cultivation of land, cattle ranching, timber production, and export. In some developed countries, incentives to expand timber production are very powerful,and tree-planting programs, which are seldom well-funded, cannot begin to compete.[30]

Agriculture provides a similar picture. The taxpayers and consumers of OECD's 24 member countries spend over \$250 billion a year on agricultural subsidies. These subsidies not only encourage farmers to expend their basic farm capital—their soils, water, and trees—they also promote overproduction. This gives rise to demands for trade protection and export subsidies to enable food to be dumped in developing countries, thus undermining their agriculture as well. Again, small, underfinanced soil and water conservation programs are too weak to compete with these subsidies.[31]

Energy provides another case in point. In many OECD countries conventional energy sources—coal, oil, and nuclear—attract enormous subsidies at various points in their development cycles from research, through exploration, construction, production, transportation, use, and waste disposal.[32] Total energy subsidies in the United States alone have been estimated at more than \$40 billion annually.[33] Again, end-of-pipe technologies to improve safety or reduce emissions, where they are available, cannot begin to compete with the opposite effects of these huge direct and indirect subsidies. Economically and ecologically perverse government interventions in the market are discussed at greater length in chapter 2.

Another condition for sustainable development concerns the nature of production. If growth rates of 3.2 to 4.7 percent are to be maintained, a significant and rapid reduction in the energy and raw material content of every unit of production will be necessary. This condition calls into question a long-favored assumption about what constitutes a "healthy" economy. Most governments, corporations, and voters assume that a healthy economy is one that uses increasing amounts of energy, materials, and resources to produce more goods, more jobs, and more income. This assumption still dominates policies in energy, agriculture, and other resource sectors. It is a holdover from the mass economy of the industrial age, which was marked by a steady expansion in the production of energy, the depletion of resources, and the degradation of the environment. During the past two decades, economic and technological changes have resulted in a leveling off of, or an absolute reduction in, the demand for energy and some basic materials per unit of production.[34] The link between growth and its impact on the environment has also been severed. A new economy has begun to emerge, one that is more efficient and potentially more sustainable, marked by people producing more goods, more jobs, and more income—but using less energy and resources for every unit of production, and more information and intelligence (see Figure 1.4).

The transformation to a more efficient economy is the result of a complex combination of factors such as new technologies and changes in historic relationships between capital, labor, and resources. Nowhere has it been more marked than in energy. Following the first oil shock, between 1973 and 1983, the 24 OECD nations improved their energy productivity on average by 1.3 percent annually. Prior to the last oil shock, when prices fell sharply, some countries, including Japan and Sweden, had reached increases in productivity of more than 2 percent per year.[35] The same trends are evident in many other areas, such as water, steel, aluminum, cement, and certain chemicals.[36]

The shift has been most evident in market economies open to change. Industries, pressed by the rising costs of energy, materials, and capital, found that they could invent products that use lighter and more durable materials and require less energy to produce. They found that they could redesign production processes to require less and more flexible capital plant and to recycle and reuse by-products internally—with benefit to their bottom line. They also found that, when they reduced the energy and material content of their products, they saved on overall costs per

FIGURE 1.4 Energy intensity and efficiency improvements of selected national economies

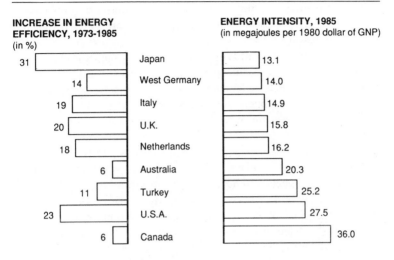

INCREASE IN ENERGY EFFICIENCY, 1973-1985 (in %)		ENERGY INTENSITY, 1985 (in megajoules per 1980 dollar of GNP)
31	Japan	13.1
14	West Germany	14.0
19	Italy	14.9
20	U.K.	15.8
18	Netherlands	16.2
6	Australia	20.3
11	Turkey	25.2
23	U.S.A.	27.5
6	Canada	36.0

Source: Compiled from International Energy Agency, Energy Conservation in IEA Countries, OECD, Paris, 1987.

unit of production *and* reduced environmental emissions and wastes. This often proved to be a far more effective way of reducing emissions than expensive end-of-pipe technologies that served no other purpose. Moreover, the environmental benefits of resource reduction and recycling extend back to the beginning of the production cycle. They manifest themselves in decreased mining and mining wastes, decreased water consumption and water pollution, and decreased air pollution, deforestation, and erosion.

Increasing the energy and resource efficiency of industrial plants or communities adds up to increasing the efficiency and competitiveness of the national economy. Those countries that have already achieved considerable progress in this direction are at the top of the international list of economic performers.[37] Between 1973 and 1984 the energy and raw material content of a unit of Japanese production dropped by 40 percent. The difference in energy intensity alone between the American and Japanese economies, for example, creates a cost advantage of the order of 5 percent for the typical Japanese export.[38] Nonmarket economies in Eastern Europe and the developing world did not share in this efficiency revolution, and many now suffer from both severe environmental degradation and serious competitive disadvantage.[39]

A rapid reduction in the energy and material content of growth is not only possible, it is essential if Western nations are to play a leading role in addressing threats to global security. Whether in satellites or stoves, truck transport or toasters, forestry or farming, society can use information and other technologies to reduce the size, cost, energy, and resource requirements of products while making them more intelligent and more environmentally sustainable. Productivity gains of 1 to 2 percent a year are quite realizable and, in the case of energy, would buy the time needed to increase the use of renewable forms of energy and to develop some new and more benign forms of power.[40]

The problem is that while technology, driven in part by market forces, has made this possible, our public and private institutions lag far behind, as do many of the key assumptions underlying the economic policies guiding these institutions. Falling oil and commodity prices have removed the pressure to increase efficiency, but they have not been replaced by compensating public policies. Industrial development grants, agricultural subsidies, water pricing, and forest and mining leases continue to encourage the extraction and use of more rather than less. Tax and fiscal incentives, pricing and marketing policies, and exchange-rate and trade-protection policies do the same. They influence the environment and resource content of growth in the wrong way. As a result, the momentum created during the period 1973–1985 has been reversed. World energy demand, for example, jumped 3.7 percent in 1988, after a 2.8 percent rise in 1987. The International Energy Agency (IEA) predicts that, if current policies are maintained, five years into the next century the world will be burning 50 percent more fossil fuels than it was in 1988. These figures are the building blocks for a global disaster. The Gulf war may drive energy prices higher or lower. Whatever the result, it can be only temporary: large fluctuations will continue, turning efficiency on and off. There is no substitute for basic reform of public policies.[41]

The environment and the economy must be integrated in our major institutions of decision-making: government, industry, and the home. This is perhaps the most important condition for sustainable development. Too often, the policies, prices, and investment decisions that drive the economy act directly against the requirements of sustainability. The environment is introduced into decision-making only after a problem has developed. At that late date, the options are usually limited to various add-on investments for end-of-pipe technologies to recapture emissions from waste streams and put them somewhere else. This leads to the still-dominant mentality that a conflict exists between a healthy economy and a healthy environment. Environment and development can obviously be mutually destructive. They can also be mutually reinforcing. If they are to be the latter, integrated decisions have to be made at the front end of the development cycle when societal goals and policies are being set, not at the tail end after society and the economy have already incurred the costs of unsustainable development. Achieving this goal will require a fundamental reorganization of economic policies and priorities, as discussed in chapter 2.

A transition to more sustainable forms of development does not mean returning the planet to a hypothetical ''natural'' state in which human activities have no impact.[42] People cannot live on the earth without altering the environment. The basic food and energy needs of 5 billion people (with 5 billion more to come in the next five decades) require large appropriations of natural resources, and the most basic aspirations for material consumption, livelihood, and health require even more. The maxim of sustainable development is not ''limits to growth''; it is ''the growth of limits.'' We must learn to recognize, and live within, limits of physical impact beyond which degradation of ecosystems, resources, and consequently human welfare are inevitable and progressive. Some limits are imposed by the impact of present technologies and social organization. But many can be expanded through changes in modes of decision-making, changes in some domestic and international policies, and massive investments in human and resource capital.

Since the WCED presented its vision of sustainable development in 1987, there has been a lively debate over the specific conditions of global sustainability. Perhaps not surprisingly, the debate has been uninformative. Knowledge of economic and environmental dynamics is still limited. Although it is possible to state the general directions in which development must proceed in order to be more rather than less sustainable—as these strategic imperatives seek to do—it is not yet possible to define the precise conditions for sustainability in respect to each specific development. Nor is it necessary. At this point, sustainability is best regarded both as a social goal and as a criterion for development. In this respect it resembles other worthy and widely accepted, but conceptually difficult social goals, such as democracy, justice, or even national security, economic development, and a healthy environment. It is possible for practical people to agree in rough terms on what each of these concepts means without agreeing on a precise definition. The same is true of the concept of sustainable development. The June 1992 Earth Summit in Rio will provide an early test. The primary purpose of the summit is to launch a global transition to sustainable development. If it is to achieve that goal, the summit will have to address a range of recommendations for policy and institutional reform in each of the areas mentioned above. These recommendations are now the subject of intense debate in a preparatory process that involves not only representatives of most governments, but also of hundreds of major industries and thousands of nongovernmental organizations.[43] The background work will be done. But it will take a Herculean effort to marshal the political resources needed for real change. The following chapters provide some measure of the size of that effort.

2

Recasting Domestic Policy

ⅬⅬⅬⅬⅬⅬⅬⅬⅬⅬⅬⅬⅬⅬⅬⅬⅬⅬⅬⅬⅬⅬⅬⅬⅬⅬⅬⅬⅬⅬ

The concept of sustainable development has fundamentally changed the nature of the debate about environment and its relationship to development. In the late 1960s, when many environmental threats began to command attention, the task was seen to be primarily one of cleaning up the mess left by two decades of rapid and unrestricted postwar growth. Environment was viewed largely as an "add-on" to development, seldom as an integral "build-in." Most professional economists saw the environmental costs of development as external to development, to be dealt with by measures external to development. This mind-set was pervasive and became reflected in add-on institutions promoting add-on policies often requiring add-on technologies. Underlying all of these add-ons—and more important than any of them—was the early but slowly changing view that environment is also an add-on politically. The environment debate thus focused mainly on the adverse impacts of development on the environment; the impacts of a degraded environment on prospects for development were largely ignored.[1] The concept of sustainable development has brought that focus into the debate.

It has changed the nature of the debate in another sense. Ever since the Club of Rome report *The Limits to Growth* was published in 1972, an important part of the environmental debate has been rooted in the assumption that environment and development are irreconcilable.[2] *The Limits to Growth* assumed a set of relationships between population, industrialization, pollution, and depletion of natural resources that led inevitably to the collapse of world order. It gave birth to a widespread movement advocating zero or even negative growth.

When the notion of reconciling environment and development was first put forward at the United Nations Conference on the Human Environment in Stockholm in 1972, it was considered revolutionary.[3] Indeed, the conference was barely able to contain the fears and suspicions of most developing and some industrialized countries that concern for the world's environment threatened their development prospects. Their watchword, taken from the experience of the rich countries in the 1950s and 1960s, was "development first and environment later, when we think we can afford it." This attitude is slowly changing as Third World leaders become aware that environmental destruction, now at a pace and scale never before seen, is undermining their prospects for economic development—and even for survival.

This change in attitude has occurred more rapidly in the richer industrialized countries. The special environmental protection and resource management agencies established during the late 1960s and early 1970s have managed some important, if limited, successes.[4] People enjoy cleaner air and water in some areas, more parks and nature reserves, and greater controls on chemicals and toxic wastes. In some countries, legislation requiring prior assessment of certain investments has been adopted and enforced. Industries have developed a wide range of clean and cost-effective technologies. Internationally, an impressive series of bilateral, regional, and global agreements and conventions has been negotiated and, in some areas, implemented.

The institutional and policy response of the 1970s—after-the-fact clean up of damage already done—was clearly necessary. But it is now seen to have been doubly inadequate. First, the limited achievements of the past two decades are to be found almost exclusively in the richer Western nations, including Japan. Eastern Europe suffered from systemic policy failure,[5] while developing countries simply were unable or unwilling to bear the costs of the expensive add-on, react-and-cure policies that characterize environmental action in the West.[6] Many developing countries, as noted, experienced a massive deterioration of their environment, with environmental problems associated with sudden industrialization and explosive urbanization being added to those associated with underdevelopment and poverty.[7] Second, even in the richer industrialized countries, progress on the first generation of environmental problems has not been great enough to catch up with the backlog, let alone keep up with the pace of development. In many areas we have fallen further behind. In the meantime, a new generation of increasingly complex environment and development issues has emerged implying potentially heavier social, economic, and political costs, whether action is taken or delayed.

Environmental issues have suddenly become deadly serious. A sea change in public opinion has forced them to the top of political agendas in the United Nations, in Washington, Tokyo, Brussels, Moscow, and other capitals around the globe, in some multilateral banks, and in many of the boardrooms of the Fortune 500.[8] The pressure on governments to do more is insistent, and has focused increasingly on calls for action that cut across traditional boundaries between public agencies and between domestic and foreign policy.

The present institutional and policy framework for environment and development is not capable of meeting this new political challenge. Environmental protection and resource management agencies, national and international, have tried to do the best job possible within the limits of their mandates and budgets, but they cannot address the sources of the problems. The agencies are small; they lack political clout; and the budgets available to them are often derisory in comparison with the tasks they have been assigned. More significantly, their mandates limit them to action on the symptoms of the problems—the negative effects of development—at the downstream end of the development cycle. The central economic and trade agencies, the development assistance agencies and banks, the energy, agriculture, and other key sectoral agencies that can influence the sources of the problems have been given no responsibility to do so. With their critical policy power and enormous budgets, these agencies were, and remain, in a position to encourage forms of development that generate negative effects on resources and the environment at rates far in excess of the capacity of environmental agencies to deal with those effects. A fundamental recasting of the system is required to do the job.

The changes required are both broad and deep, involving the entire range of government responsibility. The balance of this chapter considers several changes that are urgently needed in domestic policy. Chapters 3 and 4 focus on the international scene—chapter 3 on global environmental and geopolitical change and chapter 4 on global bargains and reform of international institutions. Many of the proposed changes may seem at first glance to be impossible, but the urgency of the sustainability question and the rapidly changing politics of environment and development prompt caution before rendering quick judgments about feasibility. As Eastern Europe demonstrates, with political change the impossible may become the inevitable. Any politician, contemplating the changing role of the environment in the economy today, could do worse than follow the Queen's advice to Alice: practice thinking of six impossible things before breakfast.

The most important changes concern: (1) the way governments intervene in the market to create incentives and disincentives for different types of economic activity; (2) the design of tax systems; (3) the counting of economic activity to measure the "health" of an economy; and (4) the ways in which corporations and the central economic, finance, trade, and sectoral agencies of government, including local government, are (or are not) made responsible or held accountable for the environmental consequences of their policies, projects, budgets, and expenditures. Western nations should lead in making the necessary changes.

Integrating Environment in Economic Decision-Making

The primary language of sustainable development is the integration of the environment and the economy. If economic activity is to result in sustainable forms of development, environment can no longer be regarded as a factor separate from and secondary to economic decision-making; it must be fully integrated into the economic decision-making process in government, industry, and the home.

In the real world, of course, the environment is fully integrated with the economy. The environment is burdened by billions of economic decisions taken daily to design, produce, market, buy, use, and dispose of goods and services. But most individuals, corporations, and governments make economic decisions without considering how those decisions will affect the environment. There are virtually no means in place to ensure that the consequences of a proposed decision are fed back implicitly or explicitly into the criteria that govern that decision before it is made. The consequences are not reflected in the price paid for the good or service in question, or in its label, marketing, or advertising. Without this feedback, individuals, corporations, and governments systematically undervalue the environmental consequences of their decisions, leading to overconsumption of resource capital and degradation of the environment.

Economic decisions reflect the incentives people face and the information they possess. This includes the formal incentives of price, revenues, profits, bonuses, and incomes. It also includes the policies and procedures of public and private bureaucracies, workplace evaluation systems, and a large class of subtle, informal incentives such as social praise or censure and shared community values, whose effectiveness is often underestimated by economists. If people's decisions are socially destructive, as their environmental decisions clearly are, then the incentives they face must be inappropriate. To improve the decisions, one must correct the incentives. Once this has been done, decisions will progressively improve as people and organizations integrate the new reward systems and information into their consciousness, habits, and decision-making procedures.

Correcting Perverse Interventions in the Market

The most important incentives are those signaled through market prices. Leaders in the Soviet Union, Eastern Europe, China, and other countries today are doing political handstands to open their economies to market forces. They have discovered that the market is the most powerful instrument available for driving development. What they haven't yet discovered is that it can drive development in two ways—sustainable and unsustainable. Whether it does one or the other is not a function of an "invisible hand," but of man-made policy.

"Normal" government interventions in the market often distort the market in ways that *actively encourage* public and private decisions that preordain unsustainable development. Tax and fiscal incentives, pricing and marketing policies, and exchange-rate and trade-protection policies all influence the energy and resource content of growth and the extent to which growth enhances or depletes stocks of ecological capital. The same is true of certain kinds of sectoral policies. Energy subsidies can, and usually do, favor large supply projects and undermine funding for biomass and renewables.[9] Tax concessions for logging, settlement, and ranching can accelerate deforestation, species loss, and soil and water degradation.[10] Pesticide subsidies can promote excessive use and thereby threaten human health, pollute water, and increase the number of pesticide-resistant species.[11] Subsidies for water-resource development and water use can lead to overuse for irrigation and industrial and municipal purposes.[12]

Agriculture provides one of the best examples of unwittingly destructive interventions in the market. Virtually the entire food cycle in North America, Western Europe, and Japan attracts huge direct or indirect subsidies, at a cost to taxpayers and consumers of over $250 billion a year. These subsidies send farmers far more powerful signals than do the small grants usually provided for soil and water conservation. They encourage farmers to occupy marginal lands and to clear forests and woodlands, make excessive use of pesticides and fertilizers, and waste underground and surface waters in irrigation. Canadian farmers alone lose well over $1 billion a year from reduced production due to erosion stemming from practices underwritten by the Canadian taxpayer.[13] Moreover, by generating vast surpluses at great economic and ecological cost, the subsidies create political pressures for still more subsidies—to increase exports, donate food as *non*emergency assistance to Third World countries, and raise trade barriers against imported food products. All of these measures hurt agricultural productivity in developing countries.

Over the next few decades the locus of global agricultural production must be shifted from developed to developing countries, where the growing demand is. Land and price reforms are helping to encourage farming in some countries in Asia, Africa, and Latin America, but these efforts could easily be undermined by the competitive dumping of Western surpluses. Third World governments are seldom able to resist shipments of subsidized or non-emergency food aid. This food relieves an ever-present and always pressing need. And it also reduces the political pressures on those governments to reform their own agricultural policies, many of which are equally perverse. Third World farmers bear most of the brunt of the resulting standoff. Even the most efficient are unable to compete with surpluses dumped at subsidized prices.[14]

Western governments should accelerate their efforts to reform policies that encourage the destruction of basic farm capital. Reducing economically perverse and ecologically destructive market distortions will require considerable structural adjustment in Western agriculture, similar in many ways to the adjustments that Western governments advocate for developing and socialist countries. However, Western countries are clearly in a much stronger position economically and politically to redesign their agricultural and related trade policies. They could redeploy their agricultural budgets in ways that encourage their farmers to adopt practices that enhance rather than deplete the soil and water base.

North American models for policies that target producers and link subsidies with conservation practices go back to the 1930s, when the Soil Conservation Service in the United States and the Prairie Farm Rehabilitation Administration in Canada brought the Dust Bowl under control.

The savings realized by phasing out existing market distortions and replacing them with conservation incentives should provide the resources needed for structural adjustment programs. They should also make it possible to restructure food "aid" to the Third World. Instead of providing nonemergency aid in the form of agricultural surpluses, Western countries should supply financial assistance in ways that encourage and support essential domestic reforms that in turn would increase production and reverse accelerating degradation of the resource base in developing countries.

In December 1990 the Brussels ministerial meeting of the Uruguay Round to reform the General Agreement on Tariffs and Trade broke down largely over agricultural subsidies. The European Communities and Japan were intransigent. Their leaders feared that if farm subsidies were cut as much as the United States and other traditional food-exporting countries demanded, the political consequences would be immediate and unpleasant. Leaders in Canada and some other countries that appeared to support reform were also concerned about the political fallout. As the negotiations broke up, some suggested that the environment, which had not been mentioned during the entire four-year round of discussions, might be the source of a breakthrough.[15] If subsidies must be paid—and few politicians are prepared to give up that certain path to power—why not pay farmers to build up rather than deplete their basic farm capital? The greening of agricultural subsidies rather than their complete elimination might be more acceptable to all parties. It could eliminate surpluses and the consequent pressure for export subsidies that distort markets and, at the same time, reduce the burden on public budgets. The suggestion may have come too late to save the Uruguay Round, but it seems clear that GATT will become increasingly preoccupied with the relationships between environment and trade. The Earth Summit in June 1992 could be an opportunity to direct GATT to take up these issues.

Forestry provides another example of destructive market intervention. The pressures on forests throughout the world vary greatly. But in both developed and developing countries these pressures are reinforced by government policies. The logging and forest industries (and clearing and road projects for agriculture and ranching) attract a wide variety of direct and indirect subsidies. The Brazilian taxpayer has been underwriting the destruction of the Amazon with millions in tax abatements for uneconomic enterprises. The Indonesians do the same. So do the Canadians. American taxpayers are subsidizing the clearing of the Tongass, the great rain forest of Alaska. Perverse incentives that encourage the overharvesting of temperate as well as tropical forests also mark world trade in forest products.[16] Western governments should review their forest policies with a view to integrating political, economic, and ecological concerns. They should ensure that the reform of forest subsidies features strongly in the negotiations about to be launched at a World Forestry Convention. They should also support the World Bank's program of sectoral adjustment loans in forestry. This policy-based lending provides support for needed changes in the policies and institutions of developing countries without which funding for separate projects cannot succeed.[17]

The transport sector, especially motor vehicles, also "benefits" from policies that are both economically and ecologically perverse. Fuel taxes in many jurisdictions, for example, still fail to distinguish between the environmental effects of different types of fuel—gas or diesel, leaded or unleaded. The tax and tariff structure, and direct and indirect subsidies, encourage heavier and more energy-intensive vehicles and road freight as opposed to rail transport in many countries. In some, private-vehicle expenses can be deducted from taxable income.[18] Western governments should examine the tax structure for motor vehicles and other forms of transport, with a view to shifting more of the social and environmental costs of transportation from society to the user.

Energy provides another example of market intervention that undermines both the earth's ecology and the world's economy. If future energy needs are to be met without putting further intolerable pressures on the world's climate, nations will have to make efficiency the cutting edge of their energy policies. As noted earlier, experience demonstrates that productivity gains of 1 to 2 percent a year are quite realizable. The major obstacle to achieving such gains is the existing framework of incentives for energy exploration, development, and consumption. These incentives are all-pervasive, they are backed by enormous budgets, and they usually promote the very opposite of what is needed for a sustainable energy future.[19] They underwrite coal, shales, oil, and gas; they ignore the costs of polluting air, land, and water; they favor inefficiency and waste; and they impose enormous burdens on already tight public budgets. In the United States alone it is estimated that total energy subsidies, including tax abatements, amount to more than $40 billion annually.[20] The figure for Canada is at least proportional, about $4 billion annually. Germany provides heavy subsidies for coal, as do China, India, and other countries. While industrialized countries have been spending billions to distort the market and consumer prices in ways that actively promote acid rain and global warming, they have been spending only a few million on measures to promote energy efficiency.[21] When these governments become serious about limiting acid rain and global warming, they will have to examine all hidden and overt subsidies and reform those that penalize conservation and end-use efficiency.[22] Leaders at the Earth Summit in Rio in 1992 should direct their representatives to take this up both within GATT and the proposed international convention on climate change.

Correcting these perverse and costly market distortions is something that free-market liberals, fiscal conservatives, budget-balancers, and environmentalists can all agree on. But correcting them will not be easy. Powerful interests derive short-term benefits from each of them and will resist any changes. This is evident from the experience with agricultural subsidies in the Uruguay Round to reform the GATT mentioned earlier. Strangely, however, in marshaling arguments for a reduction in these market distortions, governments seldom mention their negative impact on the environment and on a nation's basic stocks of ecological capital. If these factors were taken into consideration, efforts to reduce market distortions might be more successful.

Introducing Environmental Taxes and Markets

Correcting perverse market interventions is a necessary but not a sufficient condition for sustainability. Even undistorted markets are limited in several ways, the most important being that they cannot take into account the external environmental costs associated with producing, consuming, and disposing of goods and services. The market treats the resources of the atmosphere, the oceans, and the other commons as "free goods," with a zero price. It "externalizes," or transfers to the broader community, the costs of resource depletion and of air, water, land, and noise pollution. The broader community shoulders the costs in the form of damages to health, property, and ecosystems.

There is a common but regretable habit in policy discussions to presume that anything called an "externality" is, by definition, not economically important. But this is not the case: witness the depletion of the ozone layer, the potential consequences of global warming, and sea-level rise, not to mention the economic damage from acid rain, soil erosion, water degradation, and other syndromes now climbing to the top of political agendas everywhere. Environmental scarcities are growing, and habits and procedures based on zero pricing are now a clear threat to economic well-being and security.

Many critically important raw materials and goods trade at prices that fail to reflect their effects on air, water, and soil, and one way to correct this is through the introduction of environmental taxes. Properly designed environmental taxes would provide incentives for the appropriate use of certain resources. Examples of goods that should be subject to environmental taxes include fossil fuels[23], chlorofluorocarbons, electricity, plastics, pesticides, chemical fertilizers, and energy-consuming capital goods such as motor vehicles. Direct taxes on emission of pollutants and the creation of waste products requiring disposal would also be appropriate. Environmental taxes on these products would be one way to introduce "conservation pricing" as recommended by the WCED. In the case of energy, taxes would be used to influence pricing in ways that encourage conservation and end-use efficiency and discourage the use of fuels that lead to acid rain and global warming. This would involve raising energy taxes during periods of low real prices and reducing them during periods of high real prices. The objective would be to maintain prices at levels high enough to encourage steady increases in energy productivity. Stricter regulations should also require steady improvements in the efficiency of appliances and associated technologies, from electrical motors to air conditioners, in building design, automobiles, and transportation systems. And institutional innovation will be necessary to break utility supply monopolies and to reorganize the energy sector so that energy services can be sold on a competitive, least-cost basis.

There are other ways to deal with externalities. Externalities often arise from the absence of clear rights to the property in question.[24] Because of their fluidity, it is very difficult for anyone to hold proprietary rights to a particular piece of air or river. With no owner to control access and profit from the resource's preservation and responsible use, it is liable to be plundered by all and sundry.[25] In fact, that is what is happening in the atmosphere. This global commons has been exploited as a free and infinite resource, without regard to its ultimate capacity. With some ingenuity, markets can be created where none presently exist. The new market can then be left to operate largely unconstrained. Examples include systems of tradable emission permits, as is now proposed for sulfur emissions from power plants in the United States and for CO_2 emissions by nation states.[26] Other examples include water rights, wildlife ranching, and systems of deposits and refunds on hazardous or recyclable wastes.[27] By such an approach, the efficiency of markets can be harnessed to achieve efficient implementation of environmental goals articulated through political choice.

Governments should examine the feasibility of a gradual shift in the burden of taxation, reducing taxes on income, savings, and investment, and increasing them on energy and resource use, on polluting emissions to air, land, and water, and on products with a high environmental impact. This type of reform of the tax system need not increase taxes overall. Hikes in taxes on energy, resources, emissions, and waste products could be matched by an equivalent reduction in taxes on labor, savings, and investment. Taxes could then have an environmentally positive impact on consumption patterns and on the cost structure of industry without adding to the overall tax burden on industry and society. Moreover, the reduction could be graduated so that lower income groups would be compensated, in effect, for having to pay higher prices for energy and resource-intensive products. The impact on employment could even be positive. There are always formidable political barriers to new taxes but, with increased awareness, and judicious coupling of increases in environmental taxes with decreases elsewhere, the politics of changing tax and other incentive systems should not be insurmountable.

Some leaders of the most advanced industries have welcomed analyses linking the environment and the economy.[28] And a number of governments in Europe have already begun to introduce environmental taxes. In April 1990 the Swedish government introduced legislation to impose a value-added tax on energy consumption, another tax on emissions of carbon dioxide, and still another on sulphur emissions from oil, coal, and peat. Large power plants will be required to pay a charge on nitrous oxide but will get a rebate if they reduce emissions.[29] This tax reform, which may see environmental taxes applied to other products, was set to go into effect on January 1, 1991.[30] Finland moved first with a carbon tax, effective January 1, 1990. Its 1990 budget also introduced a tax on phosphate fertilizers and increased a number of existing environmental taxes. Norway imposed taxes on CFCs beginning in July 1990, and it recently increased taxes on gasoline, graduated to discriminate against lead. The Italian cabinet proposed an even more extensive package, embracing not only fossil fuel emissions but also plastic products, fertilizers, herbicides, and aircraft noise. In 1990 and 1991 Germany was debating proposals to impose a significant tax on oil and to compensate with reductions in income tax.[31] The Gulf war could lead to higher or lower oil prices. Whatever it does, the conclusion of hostilities might provide an unusual opportunity for these and other governments to raise energy taxes in order to maintain prices at low political cost.

With so much interest in environmental taxes in Europe, both the European Commission and the OECD are developing guidelines for them. Measures such as environmental taxes, conservation pricing, and environmental markets could affect a nation's trade. The trading patterns of some countries have a very high environmental and resource content, either on the import side, as in the case of Japan, or on the export side, as in the case of Canada or Australia. If a nation introduced measures to capture the cost of externalities in the price of a product traded internationally, its industry could lose out to competition from the industries of other countries. If a nation does not introduce measures to capture the cost of externalities, however, it may still lose out. Trade can shift the environmental and resource costs of production from one country or region to another and, if a nation's trade includes a high proportion of products with a large environmental and resource content, its economy can end up carrying a disproportionately high share of the environmental costs of global production. In practice, this appears to be the case for the resource-based economies of developed and developing countries.

Such measures may also cause trade conflicts. There have already been instances of international dispute over environmental measures that are claimed to be injurious to trade. There will likely be many more such disputes, as the conflict between the principles of free trade and national control over resources and environment becomes sharper. In the present Canada–U.S. dispute over Pacific fisheries, America accuses Canada of disguising an export control as a necessary measure for resource conservation. In the recent U.S.–EC dispute over American hormone-fed beef, America conceded that the European import restriction was sincerely motivated by health concerns, but argued that the health concerns are mistaken and that the measure is consequently an unjustified restriction of trade.

It was concern about the impact of environmental policies on trading patterns that led to the first attempt at international agreement to internalize the external costs of economic activity. This was the so-called Polluter Pays Principle (PPP), introduced by member countries of OECD in 1972. The PPP requires a polluting firm to pay the full costs of prevention and control measures, as determined by public authorities, to protect the environment from the pollution resulting from its activities. As agreed by the OECD in 1972, it is a principle of partial internalization of the costs of pollution. It is important mainly because it has the potential to cause the environmental costs of development to be reflected in the prices that consumers pay for goods, thereby biasing consumer choice in favor of those goods whose production, use, and disposal have the least impact on the environment.

Western nations, working through the OECD, should seek to strengthen and broaden the PPP to realize this potential.[32] Together with other nations at the Earth Summit in Rio, they should extend these efforts to the broader arena of the United Nations, the United Nations Conference on Trade and Development (UNCTAD), and GATT. If measures to capture the cost of externalities in the price of a product traded internationally were implemented in Western countries alone, they could serve to shift production of the most environmentally damaging goods from the West to the developing regions of the world. While this shift could represent a short-term economic gain to certain poor regions (recognizing that some are already ecologically bankrupt), it would be unsustainable locally and, eventually, globally. Moreover, in the case of trade in products that impact on global-scale environmental problems (such as global warming and tropical timbers), this shift could cancel some or all of the benefit created by measures already targeted on these problems. It is to be noted that there is already a growing trade, much of it illegal, in hazardous wastes in the direction of poorer developing countries.

Reforming Public Incentives

In terms of their consequences for environment and sustainability, decisions in the public sector are at least as important as those in the private sector. Public decisions create the framework of incentives for private decisions. Public decisions are also shaped by incentives, but they are of a different kind and are influenced by a variety of factors. The electoral system is vital. So are the various instruments through which the media, electorate, and opposition gain information about the government's performance. In this regard, however, the environment has always been and still remains vulnerable.

Ultimately, governments will require some comprehensive indicators on which to assess their performance on sustainability. Economic planning has growth, employment, and price indices; social planning has its demographic, income, education, and other indicators; environmental protection planning has indicators to measure changes in the quality of air, water, soil, and other resources. At the moment, however, there are no recognized indicators of sustainable development. The 1989 Group of Seven (G7) Economic Summit of the Arch in Paris asked the OECD to develop "environmental indicators," and work is proceeding, but it is not entirely clear from the communiqué whether the summiteers were seeking further and better indicators of environmental quality or indicators of sustainable development. Perhaps the differences were not explored.

Indicators of environmental quality focus attention on the back end of the development cycle, on the environmental effects of unsustainable development, and on add-on policies and technologies to reduce those effects. Indicators of sustainable development would focus attention on the front end of the development cycle: on energy, resource, chemical, and other inputs to development, and on policies to influence them. They would expose the crucial links between the environment and the economy. Energy efficiency is a good example. It is a measure of energy inputs to development (at whatever scale, national or local) and it is an indicator of both the environmental and economic performance of the economy. Other indicators have to be brought to bear, but a nation registering a steady increase in energy efficiency may be moving in the direction of more sustainable development, especially if the increases are found in the use of fossil fuels. Western nations should take the lead in developing indicators of sustainable development, working through OECD and other international forums, including the Rio summit in 1992.

Reforming Economic Accounting Systems

The single most important change in public-sector incentives, both at the political and the bureaucratic level, will involve reform of the aggregate economic performance measures in the national accounts. These accounts are one of the most important measures by which governments are publicly evaluated. Political leaders boast when GNP growth is fast, and are forced to make explanations and defend themselves when it is slow. Yet despite their established position, and their seeming authority, these accounts are only of recent vintage. And they are full of difficult measurement problems, many of which are solved by leaps of faith.

Current economic accounting systems harbor a fundamental error that would never be tolerated in the records of a private business. They have no complete capital account and no proper balance sheet. They are concerned mainly with the flow of economic activity, not the stocks of capital on which much of it is based. Communities and corporations recognize consumption of man-made capital. When stocks (e.g., buildings, plant, and equipment) depreciate, the amount is deducted from gross production. If the community or corporation wishes to remain in business, new investments are made to build up and maintain this capital.

At the moment, however, this does not apply to natural resource capital. The systems now in place everywhere ignore changes in the value and stocks of resources. Consequently, it is possible to show high levels of national or corporate income by running down environmental and resource stocks, even when this resource degradation is sure to bring declining incomes in the future. This is a certain road to receivership for a business, a nation, and the planet. National economic systems urgently need to be revised to recognize natural resources as capital and to measure changes in stocks for which economic values can be established, with decreases treated as capital consumption and increases as capital formation.

Current economic accounting systems also fail to place any value on, or they undervalue, the economic services provided by natural resources. Forests, for example, provide habitat for species and means to retain water and soils, the loss of which carries costs associated with reduced crop yields and flooding. Accounting systems do, however, place a value on expenditures made to prevent the loss of these services or to restore them after they have been destroyed. As a result, resource degradation—an Exxon *Valdez*, for example, or a Chernobyl—appears as a "public good," raising welfare, rather than as a "public bad," reducing it. Economic policy-makers, business leaders, and voters are consistently misled. Not only must this be corrected, annual losses (or gains) in the services provided by natural resources should be estimated and treated as a decrease (or increase) in income.

One recent study calculated the changes in capital stocks for timber, petroleum, and agricultural soils in Indonesia between 1971 and 1984. Adjusting for these changes, it found that GNP growth was reduced from 7 to 4 percent. The rapid GNP growth rate shown in the original national accounts was the misleading result of depleting resource stocks to boost current income unsustainably. If the lower figure had been known, it could have led to pressure for change in the government's development policy.[33] A correct national accounting system would also account for the environmental and resource content of international trade flows. Such a revision would permit a more accurate determination of whether a nation was in surplus or deficit in its trade.

Some difficulties would no doubt be encountered in introducing correct capital accounting in national accounts, particularly in determining the appropriate price to assign to nonmarketed environmental goods such as air and water quality. But these problems are of comparable difficulty to others already solved or successfully avoided in the present system of accounts. And many of the required changes will be straightforward, such as valuing the publicly held stocks of marketable natural resources (such as timber, minerals, or petroleum) or valuing environmental resources that directly yield a stream of earnings (such as agricultural soils).

Basic work on resource accounting and on mixed accounting frameworks has been done in France, Norway, Canada, the United States, and some other countries.[34] International organizations such as the OECD and the World Bank and a number of independent institutes have advanced work in this area. The U.N. Statistical Commission and Statistical Office are drafting guidelines for national bodies that wish to construct satellite and resource accounts. It seems unlikely that such changes can be incorporated into the current round of revisions to the U.N. System of National Accounts, but if the forthcoming Earth Summit in Rio provides direction, the pace of reform could be accelerated. Can these changes really make a difference? The power of numerical performance measures and incentives is well known in the management literature, particularly their power to induce managers to pursue the wrong goals if the measures are misdefined. This is precisely what the income bias in national accounts does to government behavior. Correcting it could make a large contribution to orienting public policy toward sustainability. Such a correction would also help to avoid misconceived and bitter battles between environmentalists and "developers" over how much growth must be "sacrificed-to-protect-the-environment." The question is frequently posed as how much environmental quality people are willing to pay for, but this seriously misrepresents the reality of the relationships between environment and the economy at this juncture in history. The question represents environmental quality as a consumption good, something that we can choose to have more or less of in the same way, and for the same reasons, as we would choose to have more or less vacations, automobiles, or ice cream cones. At levels of human environmental impact—for example, where environmental change means better or worse views, unpleasant smells, or more or fewer parks—this formulation may be appropriate. But in the present stage in human and environmental affairs, in which human activity is causing major degradation of productive agricultural, forest, and oceanic systems, widespread human health impacts, and the significant probability of catastrophic environmental change over the coming century, this formulation is desperately wrongheaded.

The introduction of indicators and appropriate revision of national accounting systems may be all that is required to correct public-sector economic decisions in the long term, given the dominant focus of most governments on managing economic growth. The political embarrassment involved in revealing that national economies, both in the industrial and the developing world, have been growing much more slowly than was thought—if at all—could likely be mitigated by a new government implementing the accounting changes both currently and retrospectively, thereby heaping the bulk of the resulting censure on its predecessors. But to provide appropriate sustainability incentives for smaller-scale decisions throughout the bureaucracy, other changes are also needed:

changes in discounting practices; internal review, audit and project evaluation procedures; the role of economic and environment agencies; and the basic agenda for public policy.

One area that needs reform is the practice of discounting the future in the evaluation of investments, projects, and policies involving natural resources. Discounting future benefits and costs of a proposed investment to "net present value" has long been considered a weakness of standard systems of economic analysis.[35] But what was simply a weakness in an era of small investments with limited local impacts has become potentially catastrophic in an era of massive investments, with often irreversible regional and global impacts. As noted earlier, certain investments today, when aggregated at an appropriate scale for analysis, can cause the depletion of forests on a national scale, the mass extinction of species, the steady loss of soil, the contamination of huge groundwater aquifers on which vast populations and economic infrastructure depend, acidification of the environment, and rising concentrations of greenhouse gases with consequent narrowing of the limited climatic range in which human life on earth is possible. The standard systems of investment analysis for natural resource projects in government and business ignore these irreversibilities and the consequent foreclosure of future options.

The discount rates currently used in investment analyses, commonly 10 percent or more, favor projects with high short-term benefits and long-term costs. When the costs can be externalized, the effect is to transfer them to future generations (who, of course, don't purchase and don't vote). At 10 percent, for example, a project that results in a depletion of ecological capital of $10 million in 100 years has a present value of $725. The implication is clear: the future consequences of global warming and sea-level rise or the future extinction of species, stemming from current investments, are of no significance. This is patently irrational.

The fundamental question concerns the extent to which the present generation has a responsibility to ensure that resource capital and ecological systems should be left sufficiently intact to permit future generations to enjoy continuing gains in economic welfare. There is of course no definitive answer to this question and further research is not likely to provide one. It is generally recognized that interest rates do not by some magic reflect this generation's relative valuation of the present and the future but, in the absence of anything better, the matter has been allowed to drift. Given the crucial importance of the question for sustainable development, Western governments should take the lead in eliminating the practice of discounting for investments in resource projects that involve irreversible damage, accumulation and persistence of toxics in the food chain, and transgenerational transfers of significant environmental costs.

This would correct what is perhaps the major deficiency in applying standard forms of benefit–cost analysis to proposed investments in resource development projects, and it would also improve the process of environmental assessments that a large number of countries now require. There are of course other deficiencies in the processes of environmental assessment that most countries have adopted. As a rule, with the exception of a few countries like the Netherlands, it does not apply to proposed policies and programs, only to projects and,in some areas, products. Also, with few exceptions, it is not mandatory for all major investments.

A number of recent studies have recommended that policies and programs—as well as projects, technologies, and products—should be subjected to a full prior assessment.[36] The purpose would be to ensure that the proposed policy, including proposed macroeconomic and sectoral policies, would encourage development that is sustainable against the broad criteria mentioned in chapter 1. In other words, environmental assessments should evolve into integrated assessments to determine both economic and ecological sustainability. Some countries are moving in the direction of mandatory "sustainability assessments" either as a result of deliberate policy choice or of court action. In Canada a series of court judgments has served to reduce the federal government's discretion to bypass assessments or delegate them to other levels of government. In response, the government has put forward amendments that would restore its discretion and discourage future attempts to use the courts to require an assessment.[37] Ottawa has also ignored attempts by a number of provincial governments to bypass or subvert processes of environmental assessment.

This only highlights the urgent need to build in safeguards against "sweetheart assessments." The political sources of such assessments are infinite and the term needs no elaboration. The Netherlands government has proposed the establishment of an independent body to conduct audits of environmental assessments and an International Independent Environmental Assessment Audit Commission is now being established and is undertaking pilot audits at the request of some developing countries. It is headquartered in the Hague.

Making Economic Institutions Environmentally Responsible and Accountable

In a comprehensive sense, what is needed is to make both public and private economic institutions fully accountable for the environmental consequences of their decisions. Economic and ecological systems are now totally interlocked in the real world, but they remain almost totally divorced in our institutions. During the 1960s and 1970s, governments in over 100 countries, developed and developing alike, established special environmental protection and resource management agencies. But governments failed to make their powerful central economic, trade, and sectoral agencies in any way responsible for the environmental implications of the policies they pursued, the revenues they raised, and the expenditures they made. The resulting balance of forces was and is grossly unequal.[38]

With virtually no exceptions, environment and resource management agencies are required to take development as a given. Their mandate usually begins where development ends. It embraces the *downstream* end of the development cycle, focusing on the effects of development and on ways and means to ameliorate these effects. Their policy arsenal is also very limited. They can monitor changes in the environment and resource base, undertake research, warn of impending threats, and initiate public education programs. They can plan and impose regulations. They can advocate changes in the form and content of growth but, as a rule, they have very little or no power to bring them about. They can deploy a few comparatively weak economic instruments (e.g., emission fees, rights trading, and so forth) and a few comparatively weak enforcement measures. Some can negotiate intergovernmental agreements within their mandate, a mandate that confines them largely to action aimed at after-the-fact repair of damage already done, the famous *res* of the environmental movement: *re*storing air and water quality, *re*trofitting industries and irrigation systems, *re*claiming deserts, *re*foresting lands, *re*storing natural habitats, and *re*habilitating urban areas.

The central economic and sectoral agencies of government, on the other hand, occupy their traditional fields with broad mandates, embracing the entire *upstream* end of the development cycle. Their policy arsenal includes the most powerful levers of government: monetary, fiscal, and tax policy; trade and foreign policy; and energy, agriculture, industry, and other sectoral policies. It is these agencies, through their policies and the enormous budgets they command, that have the power to determine the form and content of growth and to determine whether development will be sustainable or not. But they are seldom held accountable in these terms. If anybody is held accountable for unsustainable development, it is the environment agency—which has no formal responsibility for the fiscal, trade, and sectoral policies behind unsustainable development and very little power to change them. Hardly an enviable position.

Environmental protection and resource management agencies address a more or less "standard agenda," which has grown over the past two decades to include a long list of interrelated pollution, natural resource, and urban issues (see Appendix 3).[39] Given their mandates, they have no choice but to deal with these issues as environment issues alone, or as conservation and resource management issues, rather than as *interlocked* environment and development issues. As a result, environment policy has emerged as a limited policy field, essentially an add-on to other policy fields, with the mission largely to react to and remedy damage already done.

The central and sectoral agencies address the "priority" agenda of government—growth, employment, trade, defense, energy, agriculture, industry, foreign affairs, and so on. Given their mandates, they have no obligation to consider the impact of their policies on the environment or on stocks of resource capital. And they seldom do. When they do, as noted earlier, they usually assume implicitly that the environment and natural resources are "free" and inexhaustible, or that technology will find substitutes before resources become exhausted, or that the environment *should* subsidize the economy. Whatever political and corporate leaders say in speeches, any changes in priorities will require action at the highest levels of international organizations, government, and industry. The U.N. secretary general, heads of state, and chief executive officers of large corporations must establish clear and measurable objectives for integrating environmental and economic performance, and they must spell out the role of each agency or division in meeting those objectives.

Central economic, trade, and sectoral agencies must also be made directly responsible and accountable for ensuring that their policies and budgets encourage development that is sustainable. This will normally require legislative change in traditional mandates. In addition, environmental and resource management agencies must be given more political and financial capacity to deal with the effects of unsustainable development policies. If competing interests of short-term versus longer-term economic gain are to be reconciled at top levels of government and industry in nontraditional ways, environmental and resource management agencies must have cabinet rank comparable to that of finance ministries, full membership in key cabinet and policy committees, politically powerful and well-connected ministers, and increased budgets that reflect the heightened economic priority of their traditional areas of responsibility.

Even where all agencies are made responsible and accountable for the environmental consequences of their policies, it may be necessary and desirable to give environmental agencies some form of interagency authority to monitor and review the performance of other agencies on sustainability. Or this might be given to an independent authority similar in mandate to the General Accounting Office in the United States, or to auditors-general in some systems of government.[40] In private institutions, correcting prices through removing antisustainability subsidies and imposing environmental taxes will accomplish at least part of the required shift in priorities. Other measures to change their agenda could include environmental audits and balance sheets, or required reports of environmental performance to customers, regulators, and investors.

A clear indication of whether a government has shifted its agenda to address sustainability seriously will be its budget. A nation's annual budget establishes the framework of economic and fiscal incentives and disincentives, including all forms of taxes, within which corporate leaders, businessmen, farmers, and consumers make decisions. It is perhaps the most important environmental policy statement that any government makes in any year, because in their aggregate these decisions serve either to enhance or degrade the nation's environment and increase or reduce its stocks of ecological capital. It is also the single strongest statement of the government's real, as opposed to its rhetorical, agenda. The budget should be regarded as an environmental statement.

These changes amount to nothing less than an "alternative" agenda for government (see Appendix 4).[41] This "sustainable development" agenda should embrace all the issues thought of as simply environmental, natural resource, or urban issues and integrate them with the dominant issues of growth, development, employment, energy, trade, peace, and security. It should shift attention to the entire development cycle, viewing environmental degradation from the perspective of its common sources in the policies, programs, and budgets that drive development. It should lead to action on the sources of the issues in human activities, in the policies, attitudes, and behavior patterns underlying these activities, and in institutions and the forms and processes of decision-making. The Earth Summit in Rio must be a driving force for a major shift away from the "standard environmental agenda" adopted in Stockholm in 1972 and toward a new "sustainable development agenda" for the twenty-first century.

3

Global Environmental and Geopolitical Change

The international system is in the process of a political realignment of historic proportions. Mikhail Gorbachev's *glasnost* and *perestroika*, the turmoil in the Soviet Union and Eastern Europe, and the conflict in the Gulf are but three examples. Others include: the emergence of Japan as a new economic as well as technological superstate; the economic integration of Europe in 1992; the gradual decline of America's relative economic strength; the growing trade conflicts between the United States and Japan, the United States and Europe, and Europe and Japan; the changes in South Africa; and the slow emergence of Brazil, India, and China.

Until recently, global environmental change has been missing from discussions of the powerful forces altering the structure of world politics. Historically, these discussions have centered on political trends or on the economic trends that determine political trends. But economic trends also determine environmental trends and today, with the interlocking of the two systems, environmental trends are beginning to shape both economic *and* political trends. They are becoming a major force that could alter the very foundations of the international political system in the coming decades. Some will result in changes that are sudden and dramatic, like the Antarctic ozone hole. Others will result in changes that are slow and underreported, but nevertheless profound and irreversible, like the projected flooding of coastal areas from sea-level rise and the growing volume of environmental refugees crossing national borders. Still others will simply grow in intensity until conflict becomes unavoidable, like the deteriorating population–water ratio in parts of the Middle East and North Africa. Taken together, they will compel a rethinking of basic assumptions about current preoccupations, such as North–South relations, the unification of the Asia–Pacific region, and the world's hegemonic structure.

Why will global environmental change have such a profound impact on world politics? In this chapter, we try to answer that question. We begin by examining ways in which environmental change can lead to increased conflict. We then consider how the global ecological reach of the industrial economies has fostered interregional dependence, and how this dependence is increasing the political leverage of developing countries. Finally, we look at the need to broaden the concept of national security.

Environmental Change and Conflict

Although environmental change is of fast-growing importance, it is often treated superficially. One reason is that many environmental issues first presented themselves as small-scale pollution problems of concern mainly to local governments. Another is the tendency to leave natural resources out of the environmental calculus and to think only in terms of pollution problems. This is especially true in industrialized countries where the recent history of environmental awareness is linked to the growth of pollution. Moreover, when resource issues are included, traditional distinctions between renewable and nonrenewable resources lead to much confusion. The fact is that many so-called renewable resources (eroded soils, endangered species, 1,000-year-old tropical forests) are not renewable in any practical sense. On the other hand, many nonrenewable resources (coal, oil, and certain minerals), if not inexhaustible in an absolute sense, are inexhaustible in a practical sense because of technology, substitution, and the operation of the market. There is a third reason. Environmental issues have tended to be trivialized as "externalities," outside the mainstream agenda of government. They are treated as "boy scout" issues, advocated by high spirits out to save our planet and species. "Sunday school environmentalism," embodied in the "plant a tree and save the world" syndrome, is alive and well in the highest places, as may be seen in the U.S. budget for fiscal year 1991 and in the declaration of the 1990 Moscow Forum.[1] Environmental changes at the regional and global level are anything but trivial. They form a class of issue for which there are few precedents. They may take the form of a new or expanded resource that changes geopolitical relationships—an open Arctic, for example, with increased navigational possibilities, cutting by one half the length of the sea lanes between Japan and Europe. More often, however, they will take the form of a reduced or destroyed resource, such as a loss of national territory and shifting boundaries between neighboring states as a result of rising sea levels, increasing tensions and the potential for conflict.

Every situation is driven by its own dynamics, but environmental changes that threaten peace and security are all rooted in economic and physical realities. Even an increase in a resource provides the opportunity for conflict, because legal regimes rarely exist for allocating rights to a good that has not existed in the past. History is in fact full of examples of nations fighting to gain control of, or to stop another nation from gaining control of, raw materials, energy supplies, water supplies, sea passages, and other key environmental resources. The drive to gain or to protect access to scarce energy and other resources is often described as a major motive underlying the foreign policies of states, especially in the industrialized world: for example, the European powers carving up Africa in the nineteenth century, the German and Japanese expansion in the twentieth, and the 1991 war in the Middle East.[2] This type of conflict is likely to increase as certain resources become scarcer in face of growing demand, and as competition for them grows.

The reduction or destruction of a stock of resources provides an even likelier opportunity for conflict. When a resource is shared between nations (an international river, for example, or an offshore or ocean fishery), its degradation or depletion requires those nations to agree on reducing their claims. Formal international apportionment mechanisms are in place for some resources, but these are normally based on the historically observed range of availabilities. More commonly, there is no formal mechanism to mediate claims. Either the resource has been so abundant in the past that conflicts were uncommon, or shares were allocated by informal tradition, or there has been continuing low-level conflict.

Even within a nation, the depletion of a resource combined with a rapidly growing population can augur economic breakdown and political instability and threaten national security. The island nation of Haiti, for example, suffers some of the world's most severe erosion, down to bedrock over much of the land. Over one million "boat people," one-sixth of the population, have fled, and the U.S. Agency for International Development (USAID) attributes much of the exodus to environmental degradation.[3] El Salvador is one of the most environmentally impoverished countries in Central America, and another USAID report says that the fundamental causes of the present conflict between rebel groups and the government are as much environmental as political, stemming from the problems of resource distribution in an overcrowded land.[4] The underlying causes of political and social upheaval and military violence often include the depletion of forests, soils, and water and the incapacity of the land to support increasing population. In 1975, for example, the Ethiopian Relief and Rehabilitation Commission said that the primary cause of the famine (and the resulting mass movement of people) was not the long drought, but the accumulated effects, over decades, of resource degradation on the one hand and increasing human and animal populations on the other.[5]

Glasnost has revealed a startling pattern of ecological devastation in the Soviet Union and Central and Eastern Europe stemming from decades of political suppression, bureaucratic corruption, and economic mismanagement. Now that environmental statistics are no longer classified as state secrets, this devastation is finally being documented. During the whole of 1990, a steady stream of studies and press reports painted a picture of intense human suffering, premature death, and economic decline. At least one Hungarian out of 17 dies from environmentally induced causes. In Leipzig, life expectancy is six years less than the national average. Experts claim that many communities in the industrial belts of the old East Germany are no longer fit for human habitation. The water is unfit to drink and, far more serious, much of the soil is unfit to use because of chemical contamination. Water courses can be restored, but the restoration of soil is another matter. If possible, it will take a long time and require huge financial outlays. Fifty percent of the river water in Czechoslovakia and 95 percent in Poland is unfit for human consumption, more than half of the forests are dead or dying from acid rain, and chemicals in the food chain and human tissue exceed acceptable limits by several orders of magnitude.

Environmental neglect also threatens the region's energy supplies.

The International Energy Agency estimates that Eastern European energy consumption could increase by 50 percent between 1995 and 2005. Yet coal, the principal source of the region's energy, is also the principal source of the region's environmental devastation. Nuclear energy is also in trouble. There are two dozen Soviet-built nuclear plants in Eastern Europe, and Western experts judge that they are unsafe and decaying rapidly. Germany has decided to close down its Soviet-made plants as rapidly as possible to guard against another Chernobyl.

Cleaning up local sources of poisoned air and water and degraded soils and forests in the Soviet Union and Eastern Europe is imperative, but it will take time and money.[6] A study by the German Institute for Economic Research estimates that the cleanup will cost $200 billion over the next decade. Other estimates go much higher. More importantly, it will take human expertise and new and innovative ways to transfer technology and capital. German leaders have become concerned about the "institutional decapitation" that followed the explusion of the old East Germany leadership. Most leaders in East German industry were members of the Communist Party, and with their departure or neutralization a leadership vacuum was created that will take time to fill. The scientific infrastructure remains more or less intact, but there is virtually no capacity to undertake realistic planning that combines ecological insight with economic expertise. Western industry is consequently still hesitant to invest, slowing down the necessary process of reindustrialization. The agonizing process of structural change will also cause political strife. Many leaders are already searching for external villains, pointing to neighboring countries and demanding that they clean up their acts and halt exports of pollution.

Linkages between environmental decline and threats to national security may be subtle but they can be decisive. Lack of water, for example, is already a serious constraint on development in 80 countries, with 40 percent of the world's population.[7] Water use doubled globally in the 40 years between 1940 and 1980 and, with increasing population and irrigation, its use will double again by the year 2000, bringing growing competition and conflict between nation states. Armed conflict over access to increasingly scarce water resources has occurred or been threatened in the Middle East (the Euphrates, the Litani, and the Jordan), Asia (the Mekong and the Ganges), Africa (the Nile), and North and South America (the Rio Grande and Rio de la Plata).[8]

The Nile is a war waiting to start, a fact that the Egyptian government clearly understands.[9] Egypt's exploding population requires vastly more land and water if it is to house and feed itself, and these can be found only in the valley of the Nile. But the growing population of the Sudan is also dependent on the Nile. Egypt is also one of the countries most seriously threatened by global warming and sea-level rise. In fact, up to one fifth of Eygpt's most densely populated and productive areas could be flooded within the next century, displacing tens of millions.

When the decline of crucial environmental resources is severe, large groups of people—those who live by pastoralism, for example, or subsistence agriculture—may lose their livelihoods. Erosion and degradation of subsistence agricultural land has created millions of environmental refugees worldwide, leading to large cross-border migrations and contributing to the rapid, destabilizing flow of rural migrants into Third World cities. Africa alone has seen millions of environmental refugees every year since the 1970s, when the phrase was coined. Such large refugee flows provoke conflict between established and newcomer groups; and they provide a large reservoir of desperate people who are easily recruited into terrorist and guerrilla movements, as the record in Africa, Central America, and South America already shows.

Environmental change may contribute to international conflict in another way: the tendency of nations to attribute responsibility for environmental decline to other nations. National leaders are accustomed to thinking in this way. When stocks of a shared resource—a fishery, a river, a groundwater aquifer—collapse or are run down, the normal reaction—inspired by political self-interest, parochialism, and institutional bias—is to attribute the depletion to the greed and profligacy (or outright malice) of neighboring states.[10] Atmospheric change, global warming, and sea-level rise are driven by many different activities, and leaders are already directing the attention of their peoples to those activities found mainly in other regions. Northern leaders point to tropical deforestation; Southern leaders stress the burning of fossil fuels.[11] More generally, industrial countries often identify excessive population growth in the developing countries as the world's single largest environmental threat, while developing countries identify excessive consumption in the industrial countries. While both claims have merit, the tendency of each side to target the fault of the other may exacerbate an already tense North–South confrontation.

The Shadow Ecologies of Western Economies

Many nations, both industrial and developing, impose large burdens on the earth's environmental systems. Some do so through wealth, some through poverty; some through large and rapidly growing populations, others through high and rapidly growing levels of consumption of environmental resources per capita. The aggregate impact of any community on the environment can usefully be thought of as the product of three factors: its population, its consumption or economic activity per capita, and its material or energy flow per unit of economic activity. Population, consumption, and technological efficiency are thus all implicated in environmental degradation.[12] The total amount of carbon dioxide emitted into the atmosphere by passenger transport, for example, is equal to the product of the number of people, the amount of transport activity per person (passenger miles), and the emissions of carbon dioxide per passenger mile. A nation can therefore impose a heavy burden on the environment through any combination of high population growth, high consumption, and the inefficient use of materials and energy.

This simple formulation is complicated by one major factor: the environment and resource content of trade between nations. Economic activity today is concentrated in the world's urban/industrial regions. Few, if any, of these regions are ecologically self-contained. They breathe, drink, feed, and work on the ecological capital of their "hinterland," which also receives their accumulated wastes. At one time, the ecological hinterland of a community was confined to the areas immediately surrounding it, and that may still be true of some rural communities in developing countries. Today, however, the major urban/industrial centers of the world are locked into complex international networks for trade in goods and services of all kinds, including primary and processed energy, food, materials, and other resources. The major cities of the economically powerful Western nations constitute the nodes of these networks, enabling these nations to draw upon the ecological capital of all other nations to provide food for their populations, energy and materials for their economies, and even land, air, and water to assimilate their waste by-products. This ecological capital, which may be found thousands of miles from the regions in which it is used, forms the "shadow ecology" of an economy. The oceans, the atmosphere (climate), and other "commons" also form part of this shadow ecology. In essence, the ecological shadow of a country is the environmental resources it draws from other countries and the global commons.[13] If a nation without much geographical resilience had to do without its shadow ecology, even for a short period, its people and economy would suffocate.

Third World economies seek to be drawn into these trading networks. It is the only way they can hope to attract the investment and technologies needed to develop and trade in world markets. But participation in such networks is a two-edged sword. The economies of most developing countries (and parts of many industrialized countries) are based on their natural resources. As shown in Table 1, their soils, forests, fisheries, species, and waters make up their principal stocks of economic capital. The overexploitation and depletion of these stocks can provide developing countries with financial gains in the very short term, but can also result in a steady reduction of their economic potential over the medium and long term.[14]

Western nations heavily engaged in global sourcing should be aware of their shadow ecologies and the need to pursue policies that will sustain them. Some countries, of course, are more dependent on them than others. Japan is a case in point. Being a resource-poor country, the world's second largest economy has to import most of its energy and renewable resources as raw materials and export them as finished products. Japan's vast population also depends on access to its shadow ecology. Seventy percent of all cereal (corn, wheat, and barley) consumption and 95 percent of soybean consumption are supplied from abroad. Poor crop yields in exporting countries due to environmental degradation or bad weather could have a grave impact on Japan's food supplies. Japanese imports of roundwoods account for one-third of the world total, and more than 50 percent of Japanese consumption is supplied from abroad. Japan is facing export embargoes of tropical logs from Indonesia and the Philippines, inspired in part by the need for conservation. It has had to turn to Malaysia to assure its supply, and it is also exploring the possibility of major new sources in the Amazon. The Japanese government is becoming aware of these resource–trade connections and it is now supporting a major examination by the Pacific Economic Cooperation Council (PECC) of the sustainability of tropical forests in Southeast Asia.

Japan's vulnerability to the effects of global environmental change is not limited to the outsourcing of basic materials; it is also found in the exports of finished products. Nearly 30 percent of Japan's export

FIGURE 3.1 OECD imports from and exports to developing countries, 1989
(U.S. $ billions)

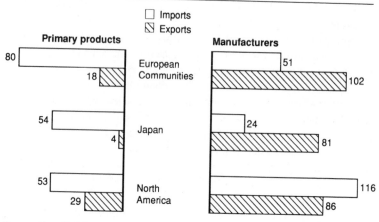

Source: Compiled from General Agreements on Tariffs and Trade, International Trade,
1979–81, 1985–87, and 1988–89; GATT, Geneva. Figures for the European
Communities are for 1987.
Exports F.O.B.; Imports C.I.F.

income is earned by the sale of automobiles, the use of which depends
on the availability of sufficient ecological space to handle their waste
emissions, especially air pollution, in the cities and towns that form
their principal markets. The fact that Japan's automobiles have been
more energy and environmentally efficient than those of its competitors
has had much to do with their success in export markets. The margin,
however, is growing smaller, partly because Japanese firms have begun
to produce and export more large, high-consumption automobiles. This
should concern the Japanese government. If environment can be used
to justify controls on imports of foodstuffs, it can also be used to justify
controls on imports of automobiles. Japan must remain sensitive to
global environmental trends, lest its exports be limited by stringent
measures to protect the "ecological space" available in foreign markets.
The same is true of other countries.

As can be seen from Figure 3.1, these dependencies are large and
growing and they apply to North America and Europe as well as to
Japan. The figures on exports and imports reveal that the dependencies
are also reciprocal. The environment and resource content of trade flows,
which is usually hidden in the trade figures, is increased by the global

reach of domestic policies, especially policies to increase trade. The agencies that make decisions on trade policies, however, are usually unaware of, and unresponsive to, their environmental connections. Trade in agriculture, energy, forestry, and a range of industrial products suffers from government interventions that are designed to distort markets in ways that protect domestic producers and give exporters an unfair advantage. But, as noted earlier, they also destroy the ecological capital on which prospects for future trade and economic prosperity depend.

Trade in tropical timbers provides a good example. Some industrialized countries impose levies that favor the importation of raw logs rather than processed products from tropical countries. They thus gain the value-added in processing the logs. As a result, the countries exporting the logs, usually developing countries, have to cut far more forest than they otherwise would in order to earn the foreign exchange that they need for development. This overexploitation now threatens the trade itself. Export revenues from tropical timber have been falling for years and are now worth about $8 billion. They are expected to continue to fall to $2 billion by the turn of the century because the resource is simply disappearing.[15] Even the industry concerned now agrees that the trade policies and domestic incentive frameworks which govern the exploitation and trade in tropical forests must be changed. In 1986 a commodity agreement was ratified that incorporated a specific commitment by governments and industry to invest in measures to sustain the growth and production of the forests being traded. It was the first international agreement that attempted to employ novel funding arrangements to address a global issue by bridging the concerns of the importing nations, mainly in the North and the exporting nations, mainly in the South. The International Tropical Timber Organization (ITTO), based in Yokohama, Japan, was established to implement the agreement, but it is much too early to say whether it will work.

The Growing Political Leverage of the South

With détente, the relative decline of the United States, and the political turmoil in the Soviet Union and Eastern Europe, some believe that a period of dangerous instability in international affairs is inevitable. Others anticipate that the instability will arise not so much from the breakup of the old East–West structure as from a growing confrontation between the North and the South. As the cold war wanes, they see a new type of power logic emerging, pitting the poor nations against the rich nations, with environmental change providing much of the leverage the South needs to get a fair deal on economic and equity questions. This mood deserves much more attention. It touches on the heart of the political importance of global environment issues.

Many developing countries believe that the second wave of environmental concern in Europe, North America, and Japan provides them with the leverage they need in bargaining for action on their mainly economic concerns. This was evident in the 1989 session of the U.N. General Assembly. Many in the Group of 77, which brings together most of the developing countries in the U.N., saw an opportunity to hold the environment hostage to the resolution of certain equity, debt, technology transfer, trade, and other economic development issues. They wanted to press for a range of concessions on development in return for a consensus on the resolution to convene a United Nations Conference on Environment and Development in Brazil in June 1992. After difficult and prolonged negotiations, they agreed to postpone their demands to a later stage in the preparatory process for the conference. The resolution anticipates a draft convention on global warming for adoption at the conference but it also anticipates that, in return, the concerns of developing countries will be actively dealt with.[16]

Developing countries are pursuing the same goals in the embryo negotiations on the proposed convention to conserve biodiversity. The vast bulk of the world's remaining reserves of biological resources exist in the tropical forests of developing countries and can be conserved only with measures that command these countries' full support. In November 1990 the United Nations Environment Program convened a meeting in Nairobi to consider a possible convention on biodiversity. Developing countries, led by Brazil, China, and India, indicated that they would withhold their participation in the convention if it did not include measures to provide them with access to the biotechnologies they need to exploit their own biological resources. Moreover, they insisted that these technologies would have to be provided on terms that were "preferential and noncommercial." The industrialized countries, with few exceptions, stated that they would resist such an undertaking. Their industries own and control the multibillion dollar biotechnologies that the developing countries want. The rapid evolution of those technologies holds tremendous promise for agriculture, medicine, and other sectors of society but that evolution, it is widely believed, depends on respect of patents and rights of intellectual property. It also depends on the companies involved retaining access to the biological resources in developing countries. It seems unlikely that agreement will be reached soon.[17]

The Intergovernmental Panel on Climate Change was established in 1989 to prepare the groundwork for the Second World Climate Conference, which took place in Geneva in November 1990. It provided another forum in which developing countries could begin to flex their muscles. At first, the IPCC looked like an apolitical scientific organization, but appearances were deceptive; its proceedings quickly became very political. This became evident at its meeting in Washington in February 1990. In his opening statement to the meeting, President George Bush made it clear that although the United States will dedicate an additional $1 billion to research, he contemplates no early commitments to reduce fossil fuel emissions and certainly no early commitments on the equity issues of concern to developing countries.[18] The representatives of developing countries, especially India, Mexico, and Brazil, made it equally clear that they expect these negotiations to address a range of fundamental issues, including economic and ecological security; preferential access to intellectual property, science, and technology; and debt, trade, aid, and North–South equity in general.

How realistic are these expectations? Can rising environmental concerns in the North translate into increased leverage for the South on certain economic and equity issues? The answers will depend on several things, including the different ways in which global environmental threats are perceived in different regions and the differences in the politics of environment and economic development in key developed and developing nations.

The participation of specific countries in a global environmental agreement is essential to the extent that those countries are capable of degrading or denying access to a resource or a shared commons in a way that threatens the economic prospects or survival of other countries. Where this is the case, history is reasonably clear. Countries can use the threat of delay or of nonparticipation in a negotiation or agreement, or of noncompliance, to secure advantages from those countries most threatened or most anxious to obtain an agreement. There are many examples of this between industrialized countries. The United States' posture vis-à-vis Canada on acid rain, and that of France and Germany vis-à-vis the Netherlands on Rhine pollution, are two prominent ongoing cases. There are also examples between developing countries. As indicated earlier, the willingness or unwillingness of states in parts of the Middle East, Africa, and Asia to act on water shortages and diversions, deforestation, or desertification has been used to influence behavior within and between those states.

Similar threats can be employed successfully by developing countries to influence the behavior of industrialized countries. If certain global problems of concern to industrialized countries are to be successfully addressed, the full participation of key developing countries is essential. The 1987 Montreal Protocol on Substances that Deplete the Ozone Layer is a case in point.[19] During the negotiations in Montreal in 1987, the major industrialized nations were unable or unwilling to deal with certain issues of crucial importance to developing countries, including those of eco-sharing, (equitable sharing of the global ecosystem), financial burden-sharing, and preferential access to intellectual property and technology. As a result, a number of developing countries, notably China and India, with a huge potential to increase CFC emissions, refused to adhere to the protocol. At a second meeting of the parties in London in June 1990, the industrialized countries not only agreed on amendments to the protocol but also, and more importantly, agreed to establish a fund of $160 million (which can be increased to $240 million if India and China adhere to the protocol) to help developing countries finance the cost of phasing into substitutes for CFCs. In response, China and India agreed to reconsider their positions. One hurdle apparently remains: an operational commitment on technology transfer. The London meeting did not resolve this, and it is too early to say whether China, India, and other developing countries will demand that it be resolved before adhering to the protocol.

The active participation of developing countries is essential to the success of several other negotiations now underway, including those on climate change. As in the case of CFCs and ozone depletion, any reductions in fossil fuel emissions by industrialized nations could soon be wiped out by increases in a few developing nations. China alone, with one of the largest reserves of coal in the world, plans 200 new coal-fired stations in the medium-term future. With this kind of negative power, countries do not need to be rich and well armed to influence the behavior of great states. The problem, as experience with the Montreal Protocol demonstrates, is not that they can prevent an agreement from being reached, but that they can refuse to sign, ratify, or implement an agreement unless and until their economic and other concerns have been addressed.

Western nations could be under increasing pressure from their own citizens to accede to the demands of developing countries if that is what it takes to secure their full participation in international environment conventions. Most Western countries have experienced a distinct greening of their political landscape. Public concern about the environment has increased steadily since the late 1960s, and during the late 1980s it climbed to the highest levels ever recorded. Environmental movements grew rapidly and increased their influence (as in Canada, the United States, and several countries in Western Europe). Environmental parties emerged in a number of countries at national, state, and local levels (as in Australia, France, Germany, Italy, Sweden, the United Kingdom, and the European Parliament). They did not win a significant vote anywhere, but their existence forced the established parties to adopt greener platforms. By 1988, political pressure had become strong enough in many countries to force leaders to undergo public baptisms as born-again environmentalists. In 1989, the Netherlands government fell because of the complex politics surrounding a series of proposed environmental initiatives.

Environmental issues also shot to the top of the agenda in most international organizations and in the boardrooms of many corporations. They became permanent features of the annual summits of the G7 economic powers, which bring together the seven largest industrial democracies (the United States, Japan, Germany, France, Great Britain, Italy, and Canada).[20] The Toronto Summit in 1988 endorsed the concept of sustainable development. Paris in 1989 was something of an education for the leaders, enabling them to begin to catch up with their own public opinion. Houston in 1990 continued that education.[21]

As the 1990s began, the debate on international environment issues had taken on all of the characteristics of a growth industry in parts of Europe, Australia, Canada, Japan, and, to a lesser extent, the United States. The economic recession in late 1990 took the peak off the environmental polls, but pollsters agreed that even if an "environmental recession" were to accompany the economic recession, it would not be of long duration and would not affect the base trend, which has been moving gradually upward since the late 1960s. Environmental values are deeply imbedded in western society and have become a political force in their own right.[22]

Historically, the politicization of global environmental issues usually begins on the national scene. The process differs widely from country to country, reflecting differences in geography as well as in economic and political systems. Several countries that are geographically "upstream" and technologically advanced, such as the United States, Japan, Germany, and the United Kingdom, find it comparatively easy to draw a distinction between issues that are domestic and those that are international. Where the latter are concerned, leaders tend to highlight the uncertainties involved and to stress the importance of greater monitoring, research, and additional data before concrete measures are adopted. While clearly necessary, this is often seen as a means of postponing the economic and political costs associated with real action. Other countries that are technologically advanced but geographically "downstream" of their neighbors, such as Canada, Norway, and the Netherlands, find it more difficult to distinguish between domestic and international issues. Domestic environmental politics easily merge into international environmental politics. With the greening of domestic politics, however, the notions of "upstream" and "downstream" are beginning to lose their significance.

Most environmental issues pass through a life cycle as shown in Figure 3.2. Political conflict is most intense at the beginning, usually after scientists or environmentalists have identified the problem, say an unsafe chemical like PCBs or asbestos, or acid rain falling over New England and eastern Canada. People demand action, the industries concerned resist it, and the government waffles. The discovery of PCBs in mothers milk, or a major accident like Three Mile Island, shifts the controversy from whether a problem really exists to what to do about it. The debate intensifies as the parties to the dispute try to avoid responsibility and transfer the costs of action to somebody else, most often the taxpayer. Implementation of react-and-cure environmental policies can have a serious impact on a specific industry or community but, once in place, public attention tapers off and the government falls into the routine of day-to-day regulation and control (see Appendix 5).

As a result of this policy life cycle, national priorities for environmental action differ among countries. Japan and the state of California lead the way in combating air pollution because they were the first to

FIGURE 3.2 The policy life cycle

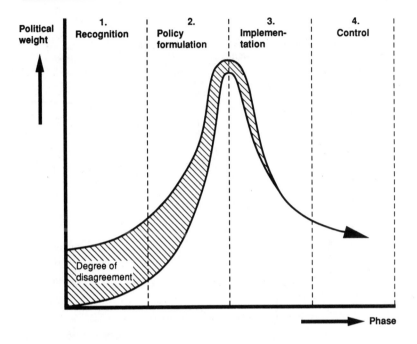

run into serious urban air-quality problems. Similarly, the United States and the Netherlands lead in the cleanup of toxic dump sites because incidents such as Love Canal and Lekkerkerk (Holland's Love Canal) forced them to act. Eastern European and developing countries are more concerned about issues that were addressed earlier by most Western countries—direct threats to human health, for instance, caused by lack of sewerage systems and contaminated drinking water and food chains. Developing countries are also concerned about issues that have never had the same priority in Western countries: the growth of deserts, the loss of forests and species, and the destruction of coastal areas. As a consequence, different countries bring different environmental agendas to the international conference table. This could, and often does, impede progress toward agreement. At the same time, however, it could facilitate bargains between self-selected groups of countries—those that earlier developed the know-how and technologies to address a problem, for example, and those that now need this expertise.

FIGURE 3.3 State-of-the-art, industrialized countries

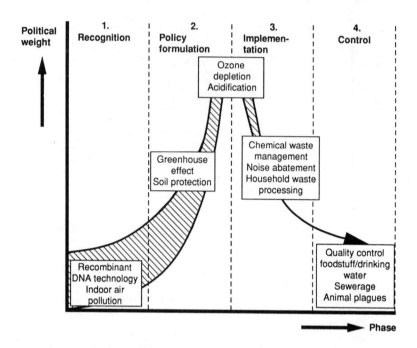

Figure 3.3 attempts to situate a number of current issues in their policy life cycle, from the perspective of Western industrialized countries as a group. Even within this group of nations, of course, priorities differ. Some issues may be in the control phase and others in the implementation phase, while in developing countries the same issues may be at a much earlier stage of recognition and action.

The general level of awareness in many developing countries is growing.[23] Among opinion and political leaders it is almost as high as in the industrialized countries.[24] But it is also different, with a much greater stress on the political imperatives of rapid economic development.[25] In many industrialized countries, environmental issues are at or near the top of the political agenda. In developing countries, the politics of environment still have a long way to go before they begin to displace traditional concerns about mainline development. As long as this difference prevails, leaders in developing countries can and will try to bargain action on environmental issues of political concern to industrialized countries against action on economic, trade, debt, and equity issues of concern to developing countries.

In fact, they have little option. As a group, developing countries may be far more dependent on resources rooted in the biosphere than most industrialized countries. They may also be more vulnerable to the effects of certain types of global environmental change than industrialized countries. But their capacity to change this and defend themselves against environmental threats can be increased only through sustainable forms of economic and social development. Moreover, large countries like China, India, and Brazil enjoy the same degree of geographical resilience to global warming, desertification, and other environmental threats as do the larger industrialized countries. What they and other developing countries lack is financial (and, perhaps, social and political) resilience. They also lack a strong infrastructure for science and technology. Budgets for monitoring and for basic and applied research are small and, in many countries, virtually nonexistent. Until this changes, developing countries will find it impossible to become full partners in international efforts to address global change. Again, that points to a strategy of rapid and sustainable economic and social development, which some of them feel is now blocked by all kinds of structural barriers controlled by the major industrialized countries.

The essential condition for the participation of any developing nation in an international convention is no different from that for an industrialized nation: it has to be convinced that the convention is in its best interest. A convention on climate change, to be successful, must not only assure developing countries that their legitimate imperative for rapid growth will not be impaired, it must also help them shift to more sustainable forms of development. It must do so at a time when the financial gap between most developing and industrial nations is widening, and when most developing countries face enormous economic pressures, both domestic and international, to overexploit their environmental resource base. A convention that does not do so may be signed, may even be ratified, but will not be implemented.

The fears of nuclear conflict that once exercised enormous power over people's minds and translated into political support for today's massive defense establishments are declining. But certain environmental threats could come to have the same power over people's minds. Both nuclear weapons and some forms of environmental degradation pose deadly threats, one comparable perhaps to a heart attack, the other to cancer. Environmental negotiations could assume some of the characteristics of arms negotiations, but they will differ in at least one important respect: the source of power of the key parties. In order to use nuclear weapons as political leverage, a country must be militarily, technologically, and economically strong. Countries need not be rich and powerful to employ environmental issues as leverage.

The Non-Proliferation Treaty (NPT) provides an interesting parallel. It prohibited those states that possessed nuclear weapons from transfering them to nonnuclear states or from transfering the know-how, technology, or materials required to manufacture nuclear weapons. It had broad support in the international community, since proliferation of nuclear weapons would pose a grave threat to the world as a whole. It was signed in London, Moscow, Washington, and 58 nonnuclear nations on July 1, 1968, and put into effect the next year. As of March 1990, 141 states had signed the treaty and 140 had ratified it, but a number of key countries with strong nuclear capabilities have done neither. They include Brazil, China, France, India, Iraq, Israel, Pakistan, and South Africa. These nations have advanced several reasons for their nonparticipation, but basically they charge that the NPT is discriminatory. It divides nations into two groups: the nuclear powers who can be "trusted" with a nuclear arsenal and who enjoy the privileges of power without restrictions; and all other nations who must accept restrictions and controls. The fact that most of the nonparticipants are Third World nations is instructive: these countries also charge that the NPT impedes their progress by imposing restrictions on the development of nuclear energy for peaceful purposes.[26]

Global environmental issues have emerged only recently, and the structure of political power on these issues has not been formed. Developing countries are raising their voices, however, and the preparations for the 1992 Earth Summit in Brazil will give them an opportunity to rethink their strategy on environmental issues as levers to achieve their development goals. The geopolitics of global environmental change have become a concern of vital importance to all countries today.

Broadening the Concept of National Security

Threats to national security have traditionally been defined largely in terms of external threats of violent aggression against peoples, communities, and the territorial integrity of states. In most countries, as a result, the security debate revolves around the narrow focus of military questions. Nonmilitary questions are weighed only when they bear directly on the military capabilities of the country, including its access to foreign sources of raw materials and energy resources that are considered vital to national well-being.[27]

Traditional perspectives on the the relationships between resources and security share a similarly narrow focus, going back to the mercantilist notion of strategic minerals as a measure of national power. Even today, the annual Posture Statement of the U.S. Joint Chiefs of Staff includes a table of mineral dependency.[28]

A broader perspective is required. The connections between resources and security are no longer limited to the classical geopolitical factors of location, land, minerals, and sea passages and other transportation corridors. As this analysis has sought to demonstrate, environmental degradation, resource depletion, and security of access to increasingly scarce reserves of energy and other raw materials may be far more important sources of human and interstate conflict in the future. This may already be evident in the southward shift of large-scale conflicts. Most of the serious conflicts during the past 45 years have taken place in the South, and the religious and racial hatreds and long-standing territorial disputes underlying many of them may well have been aggravated by environmental destruction, poverty, and economic decline. These factors could also reinforce the potential for conflict in the future.[29]

A broader concept of security, embracing environmental threats, would result in another extension of the traditional "balance of power" paradigm. This states that, in the absence of a security treaty, peace can be maintained only when contending nations or blocs maintain a balance of power between them. Following World War II, a bipolar balance of power emerged and East and West both tried to maximize their military and other capabilities to maintain their place in this balance. America's economic assistance to Europe and Japan not only displayed its supremacy, it also helped to increase the capability of its various alliances and to stimulate the international economy. Mutual dependence gradually increased as local economies were locked into those of other countries. Economic interdependence became both a fact and a precondition for national security, and a redefinition of national security became necessary. It was extended to include international economic dimensions, and the "balance of power" paradigm began to internalize the concept of economic interdependence.[30] How might an even broader concept of national security be expressed? Obviously, it would continue to assume the existence of contending nations and blocs. But the external targets that preoccupy security do not have to be limited to other states; nor do the external threats have to be limited to state-sponsored aggression. If national security were defined more broadly as the ability to counter threats to the livelihood of people and the territorial integrity and survival of nation-states, it would encompass nonmilitary threats such as environmental pollution, the collapse of life- and food-support systems, or the "invasion" of deserts and oceans.

An environmental interpretation of national security is simply an extension of the economic interpretation, since ecological capital is needed for economic development and its depletion can seriously disrupt—indeed, destroy—the economic prospects and political stability of a country or region. Because of the multiple roots of ecological interdependence and their complex relationships, even the most wealthy and powerful countries cannot shelter themselves from the consequences of change. Economic interdependence increased the dependence of developing countries on industrialized nations. Ecological interdependence has also increased dependence, although often in the reverse direction, with industrialized countries like Japan increasingly dependent on developing countries. This fact could lead some developing countries to argue against cooperation with industrialized countries on the issues of global change, unless and until they have extracted a price for that cooperation. The possibility of imposing some form of "ecological hegemony" by economic means through the back door seems remote, although some may attempt it and some will no doubt suggest that it is being attempted even where it is not.

During World War II it was recognized that new forms of international economic cooperation were needed. The result was the establishment of the International Monetary Fund (1944), the World Bank (1944), and, eventually, GATT (1948). They were followed by a new wave of regional organizations like the OECD (1960) and global organizations like UNCTAD (1964). Today, nations are linked in what the OECD has called webs of "economic and ecological interdependence,"[31] and what we have called "the interlocking of the world's economy and the earth's ecology." In an interlocked world, national security cannot be assured without new and increasingly stronger forms of international cooperation.

Global environmental change will sharply alter the present geopolitical picture. It will increase dependence among nations and regions, thereby substantially increasing the political leverage of the South. It will alter power relations among nations and it will contribute to a generally increased atmosphere of conflict, both domestically and internationally. To forestall some of the dangerous and destabilizing international risks that environmental change poses, early and broad-based international action will be needed. The next chapter discusses approaches to such action.

4

Toward Global Action

Threats to economic well-being and national security from environmental breakdown, the desire to gain or to protect access to scarce energy and other resources, and the potential collapse of life-support systems are greater today than any foreseeable threats from conventional arms. When these threats stem from military buildup by an unfriendly power, as in the Persian Gulf in August 1990, nations and regional defense blocs respond with a massive mobilization of diplomatic, military, and other resources. But to longer-term threats of environmental destruction and resource depletion, the world community seems incapable of mounting an effective response. Why?

The reasons lie partly in the fact that awareness of the scale and consequences of environmental threats are only now beginning to penetrate world councils, and partly in the complexity of the issues of global change. When these threats are considered against a background dominated by economic and ecological interdependence, and by rapid change and conflict, they emerge as complex geopolitical syndromes that challenge existing forms of governance. Measures to address them at their source confront values and aspirations, visions of the future, and codes of political behavior that are deeply rooted in our societies. They engage national and regional interests across the board in many diverse and conflicting ways. They also raise difficult questions of equitable access to the earth's ecosystem and resources, and the world's technology and wealth.

Depletion of the ozone layer, climate change, pollution of the marine environment, and similar issues are essentially global in nature. But

FIGURE 4.1 Contribution to global warming: distribution of man-made greenhouse gases, 1985

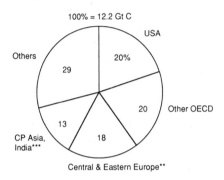

100% = 12.2 Gt C

Source: EPA; U.N.; Greenpeace; Compiled by McKinsey & Company.

* Expressed in CO_2 equivalents: higher assessment of the relative contribution of methane would raise the proportion attributed to developing nations by approximately 5% and reduce that of OECD and Eastern Europe. **Includes USSR. ***Centrally planned Asia and Indian Subcontinent (mainly Bangladesh, China, India, and Pakistan).

there is considerable room for unilateral national action, as many of the measures proposed to prevent or reduce these threats would increase macroeconomic efficiency and international competitiveness. Beyond that, however, internationally coordinated action is essential. Take climate change, arguably the most critical issue confronting the world today. Every nation is implicated, although some more than others. This is evident from Figure 4.1, which shows the current contributions of different regions to greenhouse gas (GHG) emissions, expressed in CO_2 equivalents.

The final statement of the June 1988 Toronto Conference on the Changing Atmosphere called upon the world community to stabilize atmospheric concentrations of carbon dioxide, a goal that would require a 60 to 80 percent reduction in fossil fuel combustion.[1] As an initial target, the conference proposed that nations reduce CO_2 emissions by 20 percent of 1988 levels by 2005, with the brunt of this reduction to be borne by the industrialized countries. A number of successive conferences echoed this Toronto call, with minor variations. The most important of them, the Second World Climate Conference, assembled in Geneva on October 29, 1990. After several days of scientific sessions attended by 700 of the world's leading scientists, it concluded on November 6–7 with a two-day ministerial meeting attended by heads of state and ministers and delegates from 137 countries and the European Communities.

The Intergovernmental Panel on Climate Change, which had been meeting for the previous 18 months to prepare for the conference, produced three reports that provided the point of departure for the conference. The first and most important was the "Scientific Assessment of Climate Change," produced by a committee of scientists under the chairmanship of the United Kingdom and reflecting the work of several hundred scientists from 25 countries.[2] It is the most authoritative and strongly supported statement on climate change ever made by the international scientific community and provides the scientific foundation for appropriate strategies to limit and adapt to climate change. The conference's key finding is reflected in these words: "Without action to reduce emissions, global warming is predicted to reach 2 to 5 degrees over the next century, a rate of change unprecedented in the past 10,000 years. The warming is expected to be accompanied by a sea- level rise of 65 cm ± 35 cm by the end of the next century. There remain uncertainties in predictions, particularly in regard to timing, magnitude and regional patterns of climate change."[3] In referring to the predicted range as "2 to 5 degrees," the conference rounded off the IPCC's range of 2.6 to 5.8 degrees Celsius.

The second IPCC report on the "Potential Impacts of Climate Change" was produced by a committee of scientists under the chairmanship of the Soviet Union.[4] It also marshalled the work of several hundred scientists from a large number of countries. And it confirmed and added to the work mentioned earlier, which documented the potentially catastrophic consequences of global warming and sea-level rise on freshwater resources, agriculture and food supplies, forests and species, oceans and coastal zones, cities and settlements, and national security. With this background, the ministerial conference concluded that "The potentially serious consequences of climate change, including the risk for survival in low-lying and other small island states and in some low-lying coastal and arid and semi-arid regions of the world, give sufficient reasons to begin adopting response strategies even in the face of significant uncertainties."[5]

The third IPCC report, on the "Formulation of Response Strategies," should have provided a strong basis for such strategies.[6] It didn't. It failed to pick up the conclusions from the scientific and impact assessments and follow through with clear recommendations for action by nations, groups of nations, and the international community as a whole. Chaired by the United States and dominated by diplomats, it was unable to bridge the growing gap between a majority of industrialized nations wanting to move ahead on global warming and a small minority of industrialized and Gulf states wanting to postpone action indefinitely.

The conference was the scene of intense negotiations between these two groups, the one led by the European Communities and some nations in the European Free Trade Association (EFTA), and the other by the United States and including the Soviet Union, Saudi Arabia, and the other Gulf oil states. Two days after the scientists had concluded their deliberations, the six nations of EFTA joined the 12 nations of the European Communities in agreeing "to take actions aimed at stabilizing their emissions of CO_2, or CO_2 and other greenhouse gases not controlled by the Montreal Protocol, by the year 2000 in general at the 1990 level."[7] In effect, they established a ceiling for Western Europe as a whole, a "bubble" within which the reductions by individual nations may vary but Western Europe as a whole would reduce emissions to 1990 levels by 2000. Canada, Australia, and New Zealand adopted the same target, although with some hedging. Japan finally parted company with the United States and announced a similar target,[8] although calculated on a different basis.[9] But the United States, along with the Soviet Union and the Gulf OPEC states, continued to oppose early action to reduce fossil fuel emissions of carbon dioxide.

The position of the developing countries varied greatly but the larger countries, including Brazil, China, and India, reiterated their essential position that any international agreement must address their basic concerns about equity. They will not accept legally binding commitments to reduce economic activities causing GHG emissions unless they have a clear indication that they will receive support to finance the major up-front investments needed for such action, and unless they obtain guarantees concerning the transfer of alternative technologies on a "preferential and noncommercial" basis. Some also insist on a guarantee that the funds involved are additional to normal development funding.[10]

The conference called for "negotiations on a framework convention on climate change to begin without delay," adding that "it is *highly desirable* that an *effective framework convention* on climate change, *contains appropriate commitments*, and any related instruments as might be agreed upon on the basis of consensus, be signed in Rio de Janeiro during the United Nations Conference on Environment and Development" (our italics). Given the basic differences revealed at the conference, this target date seems naively optimistic. It is highly unlikely that these differences can be overcome and an "effective" framework convention negotiated within the 20 short months between Geneva and Rio. The complexity of the issues coupled with the strong differences in national and corporate interests would seem to presage years of intense bargaining before an agreement is reached that meets even the minimal tests of fairness and political acceptability.

Governments might have a framework convention ready for signature by heads of state in June 1992 if they are willing to settle not for an "effective" but for an "empty" framework convention—nothing more than the table of contents of a more substantial agreement to be negotiated later by their successors in office. The consequences of this are discussed later.

An "effective" global convention on climate change is, of course, essential. No group of countries can expect to limit global warming on its own. Even the OECD nations taken together account for only about 40 percent of the total GHG emissions. Moreover, as noted, any reduction of the fossil fuel or CFC emissions of OECD countries could soon be wiped out by increases in a few major developing countries. For the first time in history, the nations of the world *must* cooperate. Economic and ecological interdependence has its imperatives.

The economically advanced nations of OECD bear a special responsibility for initiatives to reduce rates of climate change, to reverse deforestation and soil and species loss, and to modify ecologically perverse behavior. Apart from setting an example, their economic and trade activities are responsible historically for a disproportionate share of global environmental degradation and resource depletion, and they alone have the wealth and the scientific and institutional capacity to produce and aggressively act on a comprehensive global strategy.[11]

The main targets for policy action vary between countries. In the case of global warming, as can be seen from Figure 4.2, the targets range from fossil fuel and CFC uses in OECD nations to agriculture

FIGURE 4.2 GHG emissions by human activity: percentage of CO_2 equivalents, 1985

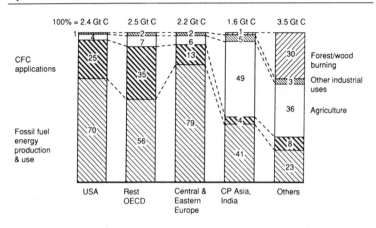

Source: McKinsey & Company.

and deforestation in developing countries. Many of the policies required to influence target groups carry short-term social and political risks. Leaders in the Third World cannot be expected to bear these risks until *after* leaders in the industrialized countries have demonstrated that they are prepared to accept them. At the moment, the performance of key Western nations does not indicate any such determination. Contrary to their political rhetoric, their policies and budgets generally encourage higher levels of fossil fuel consumption, the depletion of soils, forests, and waters, and the transfer of ecologically unsustainable technologies.

If international agreements are to be successfully negotiated and implemented, Western nations must also assure developing countries that their legitimate imperatives for rapid growth and more sustainable forms of development will not be blocked. The Toronto Conference expressed this point clearly within the context of global warming. Noting that the "growth imperative" would necessitate a range of measures, including significant additional energy use in developing countries and compensating reductions in industrialized countries, it said: "The transition to a sustainable future will require investments in energy efficiency and non-fossil energy sources. In order to ensure that these investments occur, the global community must not only halt the current net transfer of resources from developing countries, but actually reverse it."[12]

   Leadership is beginning to emerge in both industrialized and developing countries, and a number of agreements have already been concluded to address the issues of global change.[13] These may take the form of international action plans or of regional and global framework conventions. Some initiatives have been successful, but others—like the action plans to combat the spread of deserts and to provide water in Third World communities—have been paralyzed by lack of promised funds.[14] Still others are stalemated because Western nations have been unable or unwilling to address equity issues that are fundamental for developing countries.

   What scenarios for international cooperation appear to hold the most promise for real action? There are two major schools of thought. One favors the negotiation of major new conventions and protocols, possibly involving the creation of a superagency with world-scale responsibilities. The other favors a multiple-track approach, whereby negotiations on possible global conventions and protocols would be pursued, but, in the meantime, opportunities for concrete action—unilateral national action, bilateral action, and action by small self-selected groups of nations—would be seized as they arise. This school is concerned about the time required for the negotiation and implementation of meaningful conventions that could involve up to 160 independent nations. A convention has to be negotiated, then signed, then ratified, and then implemented. It could be decades before the world's climate systems begin to respond to the reductions in carbon dioxide emissions required by a convention—assuming it is enforced. We may not have decades to limit global warming, reverse net deforestation, and halt the mass extinction of plant and animal species.

   A pluralistic or opportunistic approach would proceed in parallel with the ongoing efforts to negotiate major new conventions and protocols, but would reinforce them with a number of bargains between industries, sectors, and select groups of states on a range of specific activities. It would take advantage of differences between nations and target groups to effect bargains, compacts, trades, and deals that could improve the situation of all of the parties involved. Fortunately, differences in the efficiency with which different industries, sectors, and economies use energy and other resources, differences in the technologies they deploy, and differences in the way deforestation, species loss, and other environmental threats affect different communities, all present significant opportunities for mutually advantageous bargains. Small bargains could be as numerous and diverse as the issues themselves and the nations and nongovernmental institutions involved. Less ambitious than global accords, they could address issues of immediate concern. They could provide a means of breaking out of traditional molds, gradually building trust in novel arrangements and developing the institutional and political capacity to work at larger scales. They would also enable a number of key industrialized and developed nations, as well as international organizations, to gain experience and credibility. Moreover, since they would involve a limited number of nations, they should be easier to negotiate, fund, and implement. Even informal arrangements involving nongovernmental organizations and networks could be effective in many areas. In time a large number of small bargains would enrich the contours of a broader global bargain involving most countries of the world in a united effort to ensure the quality of our environment.[15]

We now consider the potential for shaping a variety of bargains. We then take up the case of climate change: how much it could cost to reduce greenhouse gases, and the various options for funding bargains on global warming. The potential for national bargains and unilateral action will be considered before we turn to the broader question of institutions for global bargains. Some of the major opportunities for strengthening existing institutions and negotiating global conventions are considered, again with a focus on global warming.

Shaping Global Bargains

The notion of a ''global bargain'' conjures up many images, especially within the broad context of sustainable development, as discussed throughout this book.[16] In its simplest terms, a bargain involves at least two parties and two issues. It implies a trade-off between the parties on the issues. The group of nations, developed and developing, that have come together to form a bargain must agree to give up something in order to get something else. As a rule, they would give up a path of development in a given sector that is unsustainable and thus represents a threat to themselves and the other negotiating nations or the global commons. They would seek to gain the means to pursue an alternative, more sustainable, and less-threatening path of development in that sector. Any number of factors can be brought to bear to facilitate the deal:

other issues, transfer of technologies, special financing, domestic political pressure.

Bargains may involve two or more nations. Even when small in terms of the number of parties involved, they could encompass a significant proportion of an issue. Two-thirds of the world's population is contained in just 12 major geocenters: North America, the European Community, Japan, the Soviet Union, Bangladesh, Brazil, China, India, Indonesia, Mexico, Nigeria, and Pakistan.[17] Together, these geocenters control over half the world's forests, consume over 85 percent of its energy, and generate a comparable proportion of its pollution, toxic wastes, and greenhouse gases. About 40 percent of all GHG emissions come from the OECD member states, and 30 percent from Eastern Europe (including the Soviet Union), China, and India. Similarly, two-thirds of the recent destruction of tropical forests has occurred in six countries (Brazil with roughly 45 percent, followed by India, Indonesia, Burma, Thailand, and Colombia with about 5 percent each).

In considering a bargain, it is useful to have a process in mind. One is suggested in Figure 4.3. Clearly, in any process, one must begin with the objectives. For many nations the primary objective in a bargain will be environmental—making it possible for a nation to reduce the use of CFCs, for example, or to save a species-rich forest. In the case of some nations, however, the environment may be no more than a

FIGURE 4.3 Shaping a global bargain

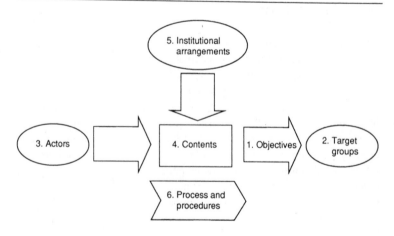

secondary or even tertiary objective. Many developing countries, for instance, may share the environmental objective, but domestic pressures stemming from their growth imperative may force them to be much more concerned about a development objective, such as debt reduction or access to certain markets or technologies. If the industrialized countries at the table are not prepared to entertain these issues and the linkages between them and environmental objectives, there may be no solid basis for a bargain. What environmental objectives might be suitable for global bargains? In considering this question, it would be useful to have an official inventory listing critical ecosystems that make up the earth's vital organs (e.g., its heart, liver, lungs), and whose immediate protection is imperative for future development, peace, and security. Even without such an inventory, it is possible to identify a number of environmental objectives suitable for bargains.

On a global scale, the ozone layer has been identified as crucial, and persuading developing and Eastern European countries to take steps to protect it could be the objective of a number of possible bargains. Similarly, measures to help countries like China and India reduce GHG emissions should be high on the list of possible bargains. Measures to reduce soil erosion and desertification in some areas and to protect certain species-rich forests are other examples of potential bargains. At the regional or local level, bargains could be aimed at regional seas like the Baltic and Mediterranean and certain coastal areas with significant feeding and breeding grounds for animal life (e.g., the Waddenzee in northwestern Europe), as well as some large bodies of fresh water (e.g., the Great Lakes), mountains and slopes essential for erosion and flood control and for maintaining river flows (e.g., parts of the Himalayas). Eastern Europe bulges with opportunities for bargains in every environmental sector, from reducing acid rain to cleaning up water supplies. Some of these opportunities have a higher value or leverage than others. As in the Third World, many ecosystems basic for development have already been partly or wholly destroyed and will have to be reestablished through massive investments in various forms of ecodevelopment.[18]

Having established the objectives, the next step in shaping a global bargain is to identify the target groups that must be induced to study their current situation, especially their expectations and needs, and to change their behavior. Most often, the environmentally perverse behavior of these groups is encouraged by domestic development policies. In previous chapters, we have considered the links between energy policies and climate change, agricultural policies and soil erosion, industrial policies and chemical pollution. We have also seen how monetary and trade policies can have an enormous impact on environmental degradation and resource depletion. In view of this, the environmental objectives of a bargain will most often depend on reform of these policies.

In fact, bargains can usefully be thought of as an exercise in negotiated policy reform. They force everyone to ask: "How much are we willing to give up in the short run to realize our environmental objectives, and what do we want in return?" Industrial countries will answer this question with an awareness of increasing domestic political pressure to "do something" about the environment. Developing countries, as noted, are under equally intense pressure to reduce poverty and foreign debt and open markets for products they can produce without further depleting their sinking reserves of ecological capital. Bargains involving East and West will reflect different pressures.

The target groups could include not only agencies of government, but also major energy consumers, heavy industries, farmers, and landowners. It will usually be necessary to consider how their interests could be served in some other way; and what changes in policies, technologies, and funding might induce them to make the changes needed to achieve the bargain's objectives. The question is not only how to persuade a government to modify a set of policies, but also how to assess and maximize the likelihood that the modified policies will result in the desired change of behavior in the target groups.

It is obviously important to bring the right combination of actors to the negotiating table. This will most often be national government officials, but could also include representatives of regional governments, private industries or utilities, and financial institutions. Two conditions must be met if a bargain is to be successful. First, each party at the table must represent, directly or indirectly, those who have the power and the resources to make the necessary changes and must be able to speak on their behalf. The actors must also be able to supply the required know-how and/or finances. Second, the negotiations must seek to establish how the changes needed can be linked so as to advance the interests of all concerned. It is here that the *differences* between the parties are all-important because it is within these differences that the opportunities for trade-offs arise. Bargains will not be implemented unless they are seen to be in the best interest of the parties. There are few "volunteers" available to lead the way in a unilateral commitment of significant resources that might weaken their competitive position or reduce their national wealth.

When a critical ecosystem has been identified, the imperatives of the target groups recognized, and the right actors gathered around the table, the actual bargain can be shaped. Again, the contents of any specific bargain will differ depending on the priorities of the contracting partners. Most bargains, however, will likely involve two elements: the transfer of technologies and know-how, and financial arrangements. The potential for innovative bargaining in both areas is considerable.

Over the past two decades, leading industries have developed an array of low and non-waste technologies and processes in just about every sector (transportation, energy, chemical, pulp and paper, agriculture, food processing, steel). Some are more productive and efficient, some provide substitutes for substances that are degrading the environment, most are economically competitive and support forms of development that are more sustainable. The industries that own these technologies and have the greatest potential for developing more productive new technologies are to be found in the Western market economies of the North. The evolving market economies of the East and the South have the greatest need for them, and also for the new technologies that can expand their "limits" to growth.

Know-how is not limited to technology, however. Many packages will have to involve the transfer of managerial and policy-making skills through training and expert support. In environmental planning, this could include assistance in policy and project assessment and in formulating and implementing crucial regulations, especially in physical planning. In other disciplines, it could include specialized contributions in land management, forestry and agriculture, industrial development (including waste management), and economic planning. Any bargain will also have to include measures to develop indigenous technological capacity.

The financial arrangements for any bargain will have to be decided by the participating countries, but the options available will be critical to the success of a negotiation. They could and should involve a major role for the private sector. In any really big deals, financial arrangements would also have to involve the multilateral banks and the trade and bilateral agencies of the developed countries concerned. Trade policies, aid policies, and debt policies will be essential components of many of these bargains, certainly the large ones involving nations in Africa and Latin America. Otherwise, they would make no sense to the participating developing countries. Given the scale of some of the issues to be addressed, international sources of financing would have to be greatly augmented. (This is discussed at some length below.) Moreover, for obvious reasons, developing countries would have to have a more decisive say in the management of these sources. All of this will require a lot of rethinking, and a few important centers and institutes have already started this process.[19]

Provided the financing and other conditions are right, the potential for a large number of small bargains is significant. Substitutes for CFCs, for example, could be offered in exchange for full participation in the Montreal Protocol (as amended), with appropriate conditions for monitoring and inspection. The technology to develop low-CO_2 reserves of fossil fuels (e.g., natural gas) could be offered against agreement not to develop more readily available high-CO_2 reserves (e.g., coal). Non-fossil fuel power generation—for instance, based on the latest solar energy technology—could be offered in exchange for agreement to forego the deployment of fossil fuel plants. Key Western and developing countries could agree to coordinate policies to reduce CO_2 emissions and to support policy changes in their own and other countries to curb deforestation. Carefully designed packages, including a mix of agricultural technologies and proposals for financing land reform in areas suitable for sustainable agriculture, could also be offered as part of an agreement to slow tropical deforestation. Some have suggested bargains negotiated on a country-by-country basis, exchanging debt relief for forest conservation.[20] Any such bargains would no doubt have to include substantially increased access to capital flows, but they could be tied to a system of incentives for policy reforms and other initiatives. In all cases, special incentive financing would be needed to make up the difference in cost to the developing country, as well as other mutually beneficial agreements on debt relief, policy change, access to markets, and the like.

The negotiation of bargains to help certain countries eliminate the use of CFCs to protect the ozone layer is one of most promising and timely strategies from an environmental point of view. At the second meeting of the parties to the Montreal Protocol in London on June 29,

1990, the delegates agreed to amend the protocol to completely phase out domestic production of CFCs by the year 2000.[21] As mentioned, a number of developing countries, with an enormous potential to increase emission levels, had not agreed to adhere to the original protocol. The parties in London, supported by the major Western nations, agreed to establish a financial mechanism, that would enable developing countries phase out CFCs and introduce substitutes. The mechanism incorporates, but is not limited to, a multilateral fund of $160 million. This can be increased to $240 million for the first three years if New Delhi and Beijing agree to adhere to the Montreal Protocol as amended—an action that the Indian and Chinese delegations in London stated that they would recommend to their governments. As of January 1991, however, neither government had signed the protocol.

With the establishment of this fund, the door is open to the negotiation of concrete bargains to phase out CFCs. Substitutes for virtually all uses of CFCs are or will be available from several multinationals. A study carried out for the Netherlands government by McKinsey & Company suggests that the annual cost of a total phaseout in developing countries (including the additional investment required and somewhat higher operating costs) would be minor in the case of gradual implementation, about $150 to $200 million per year over 5 to 10 years.[22] Subsequent studies by Metroeconomica and the United States Environmental Protection Agency suggest even lower costs. The EPA estimates the capital costs of conversion during a three-year period (1991–93)at $150–$200 million; Metroeconomica sets these costs at $200 million and expects overall costs (including operating costs) to be around $320 million.[23]

The stage for bargains on CFCs is being set in another respect. A number of Western governments have already joined with governments in some developing countries to study the feasibility and cost of eliminating the use of CFCs in those countries.[24] The United States is moving ahead with Brazil, Mexico, and Egypt. Finland is leading a study with China coordinated through the United Nations Development Program. The United Kingdom will work together with India. Canada is exploring the possibility of cooperating with Malaysia. Other countries, including Germany, Switzerland, Argentina, and Chile, have also indicated their willingness to cooperate.[25] This is very encouraging. When the basic information is available, meaningful negotiations can begin.

In discussing bargains, it is important to stress that although they may be negotiated internationally, they must be implemented nationally and locally—and it is on these levels that the greatest weaknesses in the existing institutional framework for environment and development exist. This is true in all countries but it is especially true in developing countries, which urgently need to build viable domestic institutions for sustainable development and to strengthen their scientific capacity to assess changes in the environment.[26] Future development can be sustainable only if central economic, trade, and sectoral agencies are able to assess the impact of their policies and budgets on the resource base for development, and to act on those assessments.[27]

Developing this institutional capacity should be the subject of priority bargains between multilateral and bilateral development assistance agencies and institutions in selected developing countries. Private foundations can also play a critical role by offering programs to help managers in public agencies and private industry develop the skills needed to deal with environmental issues. They could also support the development of independent policy institutes in certain countries. These bodies are often able to attract and hold better people longer than government agencies, given the history of political instability in many developing countries. Independent organizations may also benefit from association with networks of similar institutes in other countries and with bodies of international standing.

Western countries enjoy comparatively well-developed regional structures, but many of them lack both the mandate and the financial support needed to carry out more than an information exchange, or a research or advisory role in the negotiation and implementation of specific bargains. Most of the nascent bilateral and regional bodies serving developing countries are even weaker and, in several areas, nonexistent. In the area of water resources, for example, over one-third of the world's 200 major international river basins are not covered by any international agreement, and fewer than 30 have any institutional arrangements for cooperation. Most of these are found in North America and Western Europe, so the gaps are particularly acute in the developing regions of the world, which together have 144 river basins.[28] Before regional agencies can become effective partners in bargains on water resources management, species protection, halting the spread of deserts, or other syndromes, they need both the necessary mandates and the trained and experienced people working in the area of environment and development. They also need additional financial resources. Again, this could be the subject of bargains between these agencies and multilateral and bilateral development assistance agencies.

Industrialized nations should also collaborate to fund institutes to develop efficiency innovations in those sectors that are expanding the fastest in the developing world. Four or five such institutes are needed in different developing regions, each concentrating on a different industrial sector: steel, chemicals, stone and glass, and so forth. They would focus on those sectors that are mature or declining in industrialized countries, and hence are not getting much attention in efficiency research programs, but are expanding rapidly in developing countries.[29]

The Supreme Test: Global Warming

Enlisting nations to curb GHG emissions other than CFCs will be the next great challenge. Can we draw some contours of the bargains that will be necessary to induce changes in the behavior of the main target groups?

Many questions must be answered before global bargains to reduce GHG emissions can become effective. What bargaining partners should Western nations seek? Can any two countries be bargain partners? Should participation be limited to "traditional" partners in development assistance even though the opportunities for cost-effective action may be less? What about the Eastern European countries? How important are "national" bargains and unilateral action? Given answers to these questions, what measures are possible within the framework of a bargain between two countries? Should a bargain aim initially at controlling emissions, or should it settle for a more limited objective such as increasing research and improving monitoring or, as in the case of CFCs, case studies to determine the real costs and benefits to a country of reducing greenhouse gas emissions? In some instances, perhaps, where this basic information exists, bargains might focus on securing access to state-of-the-art technology or on providing financing (including swapping debt) for afforestation projects.

The answers to these questions will clearly depend on the political imperatives of the participating nations, the target groups within them, and the potential scope of a bargain. Will Western nations contemplate funding such bargains on the order of $50 million or $500 million per year? In the negotiations on a global climate convention—the Grand

Global Bargain—which commenced in Washington, D.C., in February 1991, will Western nations be prepared to establish an aggregated flow of funds amounting to $5 billion—or maybe $50 billion—per year? As a basis for comparison, UNDP resources currently fund environmental and related programs and projects to the tune of $500 million per year. And official development assistance presently totals roughly $50 billion annually.

Reducing Greenhouse Gas Emissions: How Much It Could Cost

The question of funding is of critical importance and has drawn increasingly heated debate from various sides. Still, some conclusions are becoming apparent. First, an increasing number of studies show that industrialized nations can make substantial reductions in greenhouse gas emission through energy efficiency and other measures that, at best, return a profit and, at worst, break even.[30] If the industrialized countries took advantage of these opportunities, it seems clear that they could achieve a first target—stabilization of CO_2 emissions at 1990 levels by the year 2000, for example—at a front-end cost of at most 1 percent of GNP. This is exactly the target that the nations of the European Communities, the European Free Trade Association, Canada, Australia, and Japan agreed to aim for during the Second World Climate Conference in Geneva. Front-end costs of 1 percent of GNP are significant, but few would argue that they would seriously disrupt our economy or our way of life. Moreover, as noted earlier, the efficiency and other measures purchased by this expenditure may return a net financial benefit to the industries concerned and a net economic benefit to society as a whole.

The debate in a number of countries, however, including the United States, is not centered on the costs of achieving intermediate targets. It is focussed on the costs of achieving the longer-term goal of stabilizing atmospheric concentrations of CO_2, a goal that would involve a 60 to 80 percent reduction in CO_2 emissions. This shifts the ground from the short-term costs of marginal adjustments in operational behavior to the longer-term costs of structural change. Attempts to estimate these costs have just begun. Approaches differ, but preliminary studies suggest that the gross costs (e.g., ignoring benefits) could be quite high, with estimates ranging up to 5 percent of GNP. Since some of the policies involved would aim at lowering demand through price increases, many argue that we are talking about a transition to a very different society than we now know. Without pretending to be complete, let us review some of the studies now available. We first consider the increasingly in-depth and refined studies of the short-term costs of achieving intermediate targets, then we turn to the more preliminary and necessarily speculative assessments of the longer-term costs of structural change in the area of energy and related fields.

Studies carried out in Canada, the Netherlands, and the United Kingdom suggest that many measures to reduce CO_2 emissions are economically attractive from a societal perspective on the basis of energy savings alone. Ascribing an economic value to the environmental benefits associated with these measures simply makes them more attractive. Taking as its point of departure the Toronto target referred to earlier, namely a 20 percent reduction in CO_2 emissions by 2005, a study commissioned by Canada's energy ministers found that this would involve a reduction of 291 to 342 million tons of carbon per year and that roughly 75 percent of this could be achieved through measures to improve energy efficiency. The remainder of the target would involve substituting competitive nonfossil fuel sources of electricity. An investment of C\$74 billion would yield a net benefit of C\$150 billion on the basis of energy savings alone. Investing to the limits of technical potential would reduce the net benefit to C\$99 billion.[31]

Similarly, in the preparation of a joint environmental action plan, the Dutch electricity and gas distribution companies, supported by McKinsey & Company, found that energy efficiency measures would effectively stabilize CO_2 emissions at 1990 levels by the year 2000 and at a cost of less than \$300 million a year: a 3 percent price increase.[32] In Figure 4.4, the various measures are ranked according to their net dollar gain or cost per unit of CO_2 equivalence ("carbon") conserved, and plotted on a so-called cost curve. The horizontal axis represents the absolute amount of carbon conserved, while the left-hand vertical axis denotes the net cost per unit conserved. Finally, the right-hand vertical axis gives an indication of the cumulative costs per year (in millions of U.S. dollars), provided one began with the least-cost options. Less detailed studies, encompassing overall energy consumption, suggest an even greater conservation potential that would allow the Toronto target to be realized.[33]

FIGURE 4.4 Reduction of CO_2 emissions by Dutch energy distribution sector: cost curve and cumulative costs

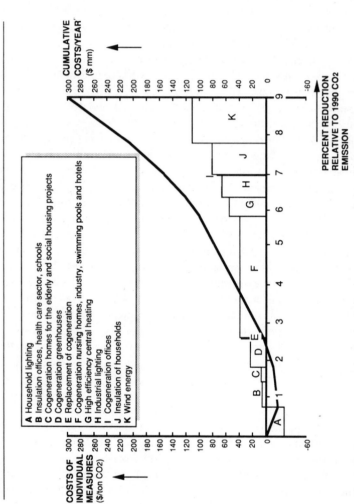

A Household lighting
B Insulation offices, health care sector, schools
C Cogeneration homes for the elderly and social housing projects
D Cogeneration greenhouses
E Replacement of cogeneration
F Cogeneration nursing homes, industry, swimming pools and hotels
G High efficiency central heating
H Industrial lighting
I Cogeneration offices
J Insulation of households
K Wind energy

Source: Environmental Action Plan, Netherlands' Energy Distribution Sector, April 1990. Compiled by McKinsey & Company.

Using a similar cost-curve approach, Tim Jackson of Lancaster University in the United Kingdom recently calculated that a combination of energy efficiency measures and high efficiency gas-fired electricity generation could reduce CO_2 emissions from the U.K. stationary sector (i.e., utilities, buildings) by 46.5 percent over current levels. If the analysis were extended to include methane emissions from fossil fuel burning, greenhouse gases emissions would be further reduced.[34]

A number of macroeconomic studies confirm these findings. Again, Europe appears to be leading the way with recent reports from, among others, Sweden, Norway, the Netherlands, and the European Community. A study carried out by the Swedish state power company, Vattenfall, and the University of Lund demonstrated the feasibility of reducing carbon dioxide emissions from the heat and power sectors in Sweden by one-third between 1987 and 2010, mainly through improvements in energy efficiency and a shift to biomass-based power generation. At the same time, nuclear power could be phased out, as required by a Swedish referendum, and the expansion of hydropower could be constrained for environmental reasons. Even so, the average cost of a unit of electricity services during this period would decrease by 18 percent. It should be noted that the scenario assumed policy changes to drive increases in energy efficiency at a rate of 1.64 percent a year.[35]

A Norwegian study, released in 1989, concluded that CO_2 emissions in the year 2000 could be stabilized at 1987 levels by increasing indirect taxes on gasoline, heating, and transport oils by 75 percent.[36] If the higher costs were offset by a reduction in other taxes (taxes on income, for instance), the macroeconomic impact would be small. Recent studies by the Dutch Central Planning Bureau point in the same direction, and those by the Commission of the European Communities go even further.[37] In a series of scenarios, the Commission showed that a "once and for all" price increase of 100 percent for coal, 40 percent for oil, and 30 percent for gas would result in a stabilization of CO_2 emissions in the year 2000 and in a 19 percent reduction in 2010, relative to 1987. This falls somewhat short of the Toronto target of a 20 percent reduction in 2005 but, once again, the macroeconomic impact is small. In fact, the models provide for an average 2.7 percent annual growth in gross domestic product (GDP) throughout the period 1990–2010, which is equal to GDP growth from 1968 to 1988. These studies no doubt weighed heavily in Europe's decision to adopt stabilization of CO_2 emissions at 1990 levels as its year 2000 target.

Studies stretching into the first decade of the next century and covering more than one country are rare. Still, to better understand the technical possibilities of addressing the global warming issue and the dimensions of the conceivable global bargains, the Netherland's Ministry of Housing, Physical Planning, and Environment commissioned McKinsey & Company to analyze the *potential* for reducing two greenhouse gases (CO_2 and CFCs) in three regions, OECD, Eastern Europe, and the rest of the world.[38] These measures apply to approximately two-thirds of all greenhouse gas emissions and they cover a sufficiently broad range of actions to provide a fair assessment of policy opportunities, but they do not pretend to be complete. Too little is known about the greenhouse gases CH_4 and N_2O to include them in the analysis.

The most important conclusion of the study, published in 1989, was that from a purely technical point of view it appears possible to reduce greenhouse gas emissions by almost 50 percent by the year 2005 relative to then-prevailing levels, and at the same time to continue economic growth at a realistic pace in each of the three target areas. This is reassuring, but some of the measures needed to achieve such a reduction are radical and would require joint action. As shown by the cost curve in Figure 4.5, however, some of them are quite profitable in the sense that they result in a net financial gain to the sector concerned, that is, the value of the energy saved more than offsets the initial investment and incremental operational costs. The first set of measures assumes complete elimination of CFC emissions (phase out, recycling), a variety of energy efficiency measures in, for example, transportation, space heating, and industrial process heat, and significant net afforestation. Implementation of these measures might eventually result in a 30 to 40 percent reduction in greenhouse gas emissions over time. Others measures impose a net cost on the sector, but may still be economically attractive to society. In order to determine how attractive, the calculus would have to place a value on the benefits of these measures; in other words, the economic damage costs prevented—damages from air pollution, acid rain, and global warming.

The cost curve shown in Figure 4.6 reveals significant potential in areas such as the phaseout of CFCs and the introduction of efficiency measures in transportation and industrial process heat, but it becomes steep when assessing measures for afforestation or the introduction of new technologies such as solar energy. It is interesting to note that the cost-effectiveness of many of these measures is greater in non-OECD countries, suggesting that bargains with these nations could be economically attractive. This was confirmed by the Chinese minister of energy at an October 1990 meeting in Beijing convened to discuss the integration of environment and economic development in China. He stated that, according to their studies, China could double its GNP without any increase in energy supply if the Chinese economy could be made as efficient as the Indian economy! The Indian economy is hardly a model, being near the bottom of the international pecking order of energy productivity. But it reflects a realistic ambition on the part of the Chinese and an enormous potential for energy and related bargains with China.

Data concerning the total financial burden worldwide to combat global warming are scarce. Preliminary estimates made by McKinsey & Company suggest a need for an international flow of funds of up to $20 to

FIGURE 4.5 Global cost curve: projection 2005

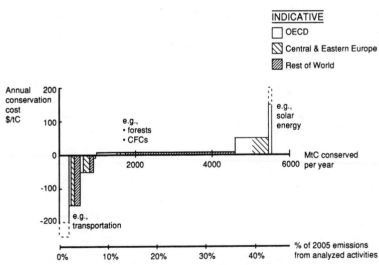

Source: International Energy Consultants. Compiled by McKinsey & Company.

FIGURE 4.6 OECD cost curve: projection 2005

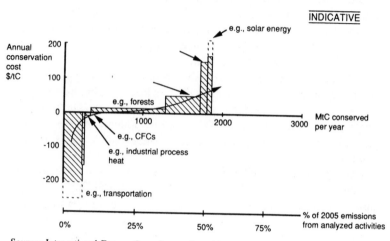

Source: International Energy Consultants. Compiled by McKinsey & Company.

$30 billion per year, assuming use of the lowest-cost measures available.[39] This is roughly one-fifth of 1 percent of the Gross World Product (GWP).[40] These amounts are not beyond reach.[41] In fact, they are less than 50 percent of the value of the subsidies that North American governments provide annually to conventional sources of energy, including fossil fuels. If this level of funding could be effectively deployed, the greenhouse gains would be very significant indeed. Forestry measures alone (both afforestation and measures to halt deforestation) might reduce emissions or absorb carbon at the rate of one-sixth of present global emissions. Energy measures in the Third World or in Eastern Europe could in theory do the same, although in practice the results would fall short. Eliminating CFCs in developing countries would reduce present levels of greenhouse gases by some 2 to 5 percent.

It is important to emphasize that each of the above studies focused on the cost of realizing certain intermediate targets. This is highly relevant from a political point of view. None of the studies suggests that the world will go bankrupt if nations take the first tangible steps needed to limit the greenhouse effect. On the contrary, the economic and environmental effects of the measures are significantly positive.[42] We can reasonably speak of "no-regrets" actions, that is, actions that would be socially, economically, and environmentally beneficial to society and should be taken even if global warming did not represent a serious threat. These studies effectively remove any excuse for delaying action.

As noted earlier, the nations of Northeastern Europe and Japan are now proceeding on this basis. At the May 1990 Regional Conference on Action for a Common Future in Bergen, Norway, organized by the government of Norway in cooperation with the United Nations Economic Commission for Europe, it became evident that in the view of most industrialized countries stabilization of CO_2 emissions at the latest by the year 2000 is a realistic objective.[43] After the Bergen conference, the IPCC's scientific assessment was leaked.[44] The government of the United Kingdom immediately switched sides, joining the action group of countries and leaving the United States in a remarkably (and—given its earlier record of environmental leadership—embarrassingly) isolated position, a position its performance at the Houston Group of Seven Economic Summit and the Geneva Climate Conference suggests that it will retain for some time.[45]

In Geneva, as noted above, the nations of the European Communities and the European Free Trade Association, as well as Japan, Canada, Australia, and New Zealand, agreed to stabilize their emissions at about 1990 levels by the year 2000. Germany has gone further: in June 1990 the then West German government committed itself to secure a reduction of 25 percent in CO_2 emissions from present 1990 levels by 2000, 5 percent more than the Toronto target. With unification and the extremely low energy efficiency of the East German economy, the new German government found that it could do even more at a lower front-end cost. Just before the Second World Climate Conference in Geneva, it raised its 2005 target to 30 percent. Some other countries are expected to follow the German lead. The Danish government has committed itself to achieve the Toronto target. The Netherlands, Sweden, Norway, France, and Italy are all expected to announce a year 2005 target during the course of negotiations on a global convention.

The above studies on the costs of realizing certain intermediate targets don't deal with the question of structural equilibrium, a fact that appears to be of particular concern to the United States. Ideally, in the longer term, the market forces of supply and demand for energy should provide the basis for an equilibrium of CO_2 concentrations in the atmosphere in line with "nonexcessive" increases in temperature. But this would require market interventions of a different kind than governments now make and the nature of these interventions is a source of much of the concern. The possible effect of these interventions on "business-as-usual" is even more worrisome to some governments.

The hard fact is that any limitation of global warming requires the generation of less CO_2, which means burning less coal, oil, and gas. This will require a major revolution in our current way of life. It could also imply a major revolution in the economic and financial world because the measures required will affect oil prices, energy taxes, inflation, and major flows of trade and revenues between oil/coal producing and consuming countries. Countries such as Saudi Arabia and Australia on the one hand, and Japan on the other—not to mention U.S. states such as Texas and Canadian provinces such as Alberta—will bring quite different perspectives to the issue. Apart from energy efficiency and renewables, phasing out of fossil fuels might involve increased development of hydro and nuclear resources in some countries, opening up another Pandora's box of heated debate.

As yet, our knowledge with regard to the longer-term structural implications of these measures is limited mostly to the realm of speculation. In 1989, Alan Manne of Stanford University and Richard Richels of the Electric Power Research Institute published the first major study of the macroeconomic impact of structural changes to limit CO_2 emissions in the United States.[46] They investigated the costs of restricting carbon emissions to the 1990 level through 2000, reducing them gradually to 80 percent of this level by 2020 and stabilizing them thereafter. In the absence of any carbon constraint, they found that emissions would increase sixfold between 1990 and 2100. They found that the effects of a carbon constraint would not begin to have measurable macroeconomic consequences until 2010. At that point, the required increase in energy prices begins to have a significant effect upon the share of gross output available for current consumption. By 2030 the total output available for current consumption is reduced by 5 percent, a percentage that remains relatively constant until 2100. Taking the whole period from 1990 to 2100, the present value of these "losses" would be $3.6 trillion, discounting to 1990 at 5 percent per year.

Manne and Richels also found that emissions could be reduced to the desired levels by a variety of policy instruments, including a carbon tax. A low carbon tax of $29 per ton of carbon would serve until the year 2000, bringing in approximately $40 billion per year (about two-thirds of the amount of subsidies that North American governments now grant to conventional sources of energy). Thereafter, the tax would have to rise sharply to discourage consumer demand as emission limits are tightened. By the middle of the twenty-first century, sufficient additional noncarbon energy capacity should be available to stabilize the tax at about $250 per ton of carbon.

It should be stressed that these figures represent Manne and Richels' most constrained scenario from the perspective of supply enhancement and price-driven demand reduction. Indeed, the researchers assumed that the rate of autonomous (non-price-driven) energy efficiency improvement during the period would be zero! Through a number of other scenarios involving, for example, the development of a CO_2 removal and disposal capability, or a transition to low-cost noncarbon sources of electricity, they found that the macroeconomic impacts could be greatly reduced. With energy efficiency improvements of 1 percent a year, U.S. energy demand by 2050 would be cut in half, and the total discounted output available for current consumption would fall to $1.8 trillion. Another scenario brings it down to $0.8 trillion, or 1 percent of annual economic consumption.

The debate on the macroeconomic impact of limiting CO_2 emissions has clearly just begun. Robert Williams of Princeton University maintains that the technological scenario assumed by Manne and Richels is too restrictive. Williams states that "the potential for energy efficiency improvements is large even with existing technology, despite the remarkable progress made in uncoupling energy and economic growth between 1973 and 1986," and that the total cost to society could be considerably lower.[47] Incorporating the potential benefits of a CFC phaseout and an afforestation program, William Nordhaus of Yale University pegs the cost of containing GHG emissions to 1990 levels at between 1 and 2 percent of national income by the middle of the next century. Assuming that the U.S. data applies worldwide, Nordhaus estimates that (at the 1989 level of world economic activity) the total cost of reducing GHG emissions by 40 percent in an efficient global program will amount to less than $100 billion. Any additional effort, however, would result in a steep increase in the overall cost: a 60 percent reduction, for example, would require over $300 billion.[48] Referring to similar estimates in a study in progress by the U.S. Congressional Budget Office, *New York Times* reporter Peter Passell notes that high costs alone need not rule out action if the alternative is catastrophe.[49]

Funding Bargains on Global Warming

In light of these figures, marshaling sufficient investment for bargains on global warming will require new initiatives. In 1990, the World Bank established a Global Environmental Facility in cooperation with UNDP and UNEP. This is an encouraging start, especially if the facility reaches the level of $1 billion annually that some of its supporters have proposed. It could serve to finance some of the more pressing immediate needs, but it would not begin to satisfy the requirements of the wide range of potential bargains discussed earlier, nor of the conventions and protocols now being negotiated. Nor is it intended to do so.[50]

As noted earlier, governments meeting in London in June 1990 to amend the Montreal Protocol on the Ozone Layer agreed to establish a fund to assist developing countries finance the incremental costs of phasing into substitutes for CFCs. The fund can draw on $160 million now and $250 million when China and India adhere to the protocol.[51]

Funds to finance monitoring, research, and major development projects in developing countries, including the transfer of technologies, will be an essential component of all future conventions. This is particularly true of the conventions on climate change as well as the conventions on biodiversity and forestry, on which negotiations have been launched.[52] Indeed, as can be seen from the studies discussed above, the requirements for funding projects and technology transfer under these conventions will be substantially higher and will extend over much longer periods of time than the requirements under the Montreal Protocol.

In anticipation of these requirements, a number of special funds have already been proposed. In June 1988 for example, the Toronto Conference suggested the creation of a World Atmosphere Fund to be financed by a levy on the fossil fuel consumption of industrialized countries, with the proceeds channeled to developing countries to help them limit and adapt to the consequences of global warming.

A number of other conferences have endorsed this idea with variations.[53] In September 1989 then Prime Minister Rajiv Gandhi of India proposed a Planet Protection Fund to be administered by the United Nations and used to underwrite the development of environmental technologies for participating nations. In submitting it to the Commonwealth prime ministers meeting in Kuala Lumpur, he suggested that the fund should be about $15 billion, raised through a levy on the GNP of all participating members. His initiative failed largely because of the opposition of then Prime Minister Margaret Thatcher of the United Kingdom. In Houston in July 1990, the Group of Seven debated another proposal. Chancellor Helmut Kohl of Germany called for the establishment of a fund of $2.5 billion to combat global warming. The United States stood alone in opposing this initiative and it failed.[54] With six of the G7 now prepared to support such a fund, however, the proposal will undoubtedly resurface in various forms during the negotiations on a climate convention and in the preparations for the June 1992 Earth Summit in Brazil.

Various means of financing the funds have also been proposed. Most would rely on national contributions based on a fixed percentage of the contributor's GNP. Some, such as the World Atmosphere Fund proposed in Toronto, would be financed through a "climate protection tax" related to the carbon content of fuels. Others, funding a wider range of activities, would be financed through levies on a mix of pollutants. A tax of 0.5 percent on the value of international trade has been suggested, which would yield about $7 billion annually. Egypt has proposed a levy of 1 percent on international passenger and freight transport, which would provide approximately $250 million per year, with a growth of 10 to 15 percent per annum. Other proposals would place a small levy on the commercial value of ocean fish catches, toxic incineration at sea, and river-borne wastes. As mentioned in chapter 2, however, it is a tax on carbon, sulphur, nitrous oxides, and other air pollutants that seems to be attracting the most attention at the moment.

A 1989 study of conservation financing options, led by the World Resources Institute (WRI) on behalf of UNDP and other sponsors, assessed present financing programs and the scope for improving them, as well as a number of new initiatives.[55] It recommended an "international environmental facility" to generate well-designed conservation programs by arranging preinvestment funding and helping to arrange cofinancing, including guarantees, for overall project packages from a variety of existing sources. It would be a joint venture of existing multilateral and bilateral agencies, including the development banks, and would have a target of $3 billion in projects (bargains) for the first five years. The WRI study also recommended a pilot investment program for sustainable resource use, called "Ecovest." This new entity would bring the private sector into potential bargains by providing intermediation services similar to those provided by an investment bank, but for conservation financing. It would "gather long-term capital, spread risks, arrange access to technology, and improve incentives for investments in such activities as wildlife utilization, sustainable forest management, the development of forest products other than timber, sustainable marine aquaculture, and so on."[56] A reasonable target for Ecovest might be $25 to $75 million, which the WRI study felt could be invested effectively over three to five years.

In 1988 vice presidents of the Royal Bank of Canada and a Montreal investment firm called for new forms of financing through the private sector.[57] They made a number of suggestions for study: "environmental mutual funds," which would include selected companies that represent environmentally progressive and sound investments; "tax-free environmental bonds" to raise money for investments to protect, restore, and enhance environmental capital; "environmental flow-through shares" to provide incentives for individuals to invest in companies that exceed environmental standards; "environmental venture-capital funds," with tax incentives to enhance investment in environmentally attractive technologies; and an "emergency environmental fund," financed through a one-time levy on corporate profits to clean up old environmental messes.

Another idea being discussed would involve offsetting increases in fossil fuel burning with increased forest cover, which acts as a carbon sink while growing. Major fossil fuel consumers, like power utilities, may wish to consider supporting afforestation projects to offset the increased carbon load they would place on the atmosphere. Adding an "afforestation surcharge" to the price of electricity, steel, aluminum, or automobiles would internalize a part of the environmental costs of global warming. In the United States, Applied Energy Services, a utility building a new coal-fired power plant, contributed $2 million to a $14 million afforestation project in Guatemala to plant 52 million trees that will sequester the same amount of carbon as the new plant will emit over its entire life. Similar initiatives are being considered in both Japan and Europe.

The need to secure foreign currency to finance debt has exacerbated resource degradation in many developing countries and policy-makers are now searching for innovative approaches that can link debt reduction with investments to preserve or increase a nation's basic stocks of ecological capital. Debt-for-nature swaps constitute a new conservation financing instrument suitable for certain types of bargains. While each must be instituted as a separate transaction for a specific activity (e.g., a reforestation project, a national park acquisition or development), they all take advantage of the spread between the market value of government bonds and their face value in local currency to provide much needed revenue for local activities. For example, the Swedish International Development Agency, in exchange for an expenditure of about $4 million to purchase Costa Rican debt, provided the government of Costa Rica with more than $20 million in local currency used to create the Guanacaste National Park. A number of deals have been negotiated involving Bolivia, Costa Rica, Ecuador, and the Philippines. The total amount swapped has been small (about $100 million) in relation to the developing world's debt burden (about $1.3 trillion), but the WRI study mentioned above encourages aid agencies to increase funding available for purchasing discounted debt to support more and larger debt-for-nature schemes.[58]

Some measures would not require governments to raise revenue directly. Several proposals have been made, for example, to issue internationally tradable permits to emit greenhouse gases and to auction off some or all of the first issue.[59] In 1989, Michael Grubb, of the Royal Institute of International Affairs in London, suggested that an international regime be created within which shares or permits to emit carbon would be granted to national governments and could be marketed freely between them.[60] In accordance with this plan, governments would negotiate a target global level of carbon emissions reflecting the capacity of the atmosphere and the dangers of global warming. Initially, permits would be granted to nations according to a per capita formula based on their population. (If only adults were counted, this would not be an incentive for population growth.) Nations that exceeded their quota would incur fiscal penalties. If quotas were large enough and enforceable, the industrialized countries, with generally larger carbon emissions per capita, would be induced to lease part of the quotas of developing countries and the scheme would thus result in a transfer of wealth from richer to the poorer countries. To avoid hoarding by the rich countries, permits would be leased and could even be reissued periodically according to the initial allocation system. In a free market, with a cap on total emissions, the marginal cost of a carbon emission right could become very high and thus provide an incentive for energy efficiency measures. The "currency" of trading could be restricted to development projects related to reducing carbon emissions.

While some of the above options are attractive, the simplest solution for politicians in the Western nations is undoubtedly to channel a fixed percentage of national income into a fund. The government of Norway has proposed that industrial nations pledge 0.1 percent of their GNP toward the promotion of greenhouse bargains with developing countries, and has offered to do so conditional on a majority of OECD countries following suit. The Netherlands' budget for fiscal year 1989 included provision for an annual contribution of 250 million guilders for similar purposes. This politically easy solution may only be acceptable, however, if both parties agree that the money be spent to improve sustainability, combining economic development with environmental protection. A number of Third World countries have expressed concern about this fearing that tests of sustainability could become a new form of "conditionality." They have also stipulated that any funds for environmental protection should be additional to the funding for development that they would have otherwise received. This question of "additionality" is difficult for want of a base figure and any reasonable method of accounting. Developing countries tend to start counting funds as "additional" when the total exceeds the official development assistance (ODA) target of 0.7 percent of GNP established by a United Nations resolution.[61] Most countries still fall far short of the 0.7 percent. In fact, on average, Western countries achieve only about half that target.[62] The political basis for North–South bargains would become very weak indeed if Western nations tried to rechannel part of the current ODA flows in a West–East direction, as was proposed by a number of countries following the collapse of the Berlin Wall.

A choice of options is not easy. A carbon tax on fossil fuels and/or a levy on a mix of pollutants are clearly favored from an environmental policy point of view, and a number of nations are considering them. To the extent that such taxes increase macroeconomic efficiency and international competitiveness, their unilateral imposition can be justified. In Canada, the recently launched debate on a carbon tax, and environmental taxes generally, has already generated fears that these levies could affect the nation's competitive position in the evolving

North American common market. While the question has been raised in the United States, it seems evident that a carbon levy, or any other kind of pollution levy, must fail the "read my lips" test of U.S. tax policy, although that test now seems to be evolving. In spite of the policy advantages of carbon taxes, applying them internationally will be difficult because of the number of activities affected, the problems of exchange-rate manipulation, and the necessity of gaining worldwide political acceptance.

Systems of tradable permits avoid many of the political problems associated with imposing taxes or arbitrary constraints on a voting population. They also capture the benefits of decentralized, market-based decisions, tending in the direction of least-cost solutions. But, even in today's world, they run into a number of ideological barriers. Barring some breakthrough, many governments will be driven in the direction of using a fixed percentage of GNP. The record proves that this is a most unreliable source, development assistance being the first or second victim of budget cuts during periods of restraint. It raises the question of conditionality and additionality. And, being far removed from the Polluter Pays Principle adopted by OECD almost 20 years ago, it provides no pricing pressure to reduce pollution within the funding countries. Unfortunately, these very failings make it politically attractive and it seems probable that a number of countries will focus on the percentage of GNP option. It may be that within blocs like the European Communities and the European Free Trade Association, agreement could be reached to raise a fixed percentage through a specific levy. Elsewhere, it might come out of general tax revenues.

During the negotiations now underway, Western nations and their Southern and Eastern counterparts will be searching for measures that both increase macroeconomic efficiency and reduce GHG emissions. In other words, they will be looking for opportunities to maximize the environmental yield per dollar spent or, conversely, to minimize the overall cost to society of stabilizing GHG concentrations in the atmosphere. Some estimates suggest that a multilateral "clearinghouse mechanism" could increase the effectiveness of any funds established by approximately 30 percent.[63] To illustrate this, consider a bargain between three countries: A,B, and C. If the cost per unit of carbon that is conserved in country A is 20 whereas in country B it is 10, A can pay B 15 to take action, and both have a financial advantage of 5. However, if the clearinghouse can locate a country C with a cost of 6, the benefit to A and C goes up to 7, and even B should even pay C to induce action. Before such a mechanism could become operational, countries would have to agree on reduction targets and the financial contribution expected of each them.

National Bargains and Unilateral Action

Many governments tend to reject the notion of small bargains involving just a few countries, and of unilateral national action. They advance various arguments to support this position: uncertainty; the fact that no one nation, even a large one, can solve the problem acting on its own; and the fact that unilateral action may encourage "free riders" and may jeopardize a country's international competitive position. Most of all, they seek the comfort of an international agreement in which the political misery will at least be abated, if not shared equally.

Scientific uncertainty, of course, is endemic. Environmental issues, like most other issues—the economy, trade questions, foreign policy, military action—seldom come wrapped in certainty. But that, in itself, is not a reason for inaction. Given the potentially catastrophic consequences of ozone depletion, global warming, species loss, and other syndromes, the cost of insurance against them is not large. The uncertainties surrounding military security are much greater and nations spend colossal sums to buy insurance in the form of armed forces, matériel, and highly unpredictable technologies. Environmental insurance is especially cheap when you consider that the most cost-effective measures against global warming are also the most cost-effective ways to deal with acid rain, deforestation, and other issues of more immediate concern. Moreover, up to a certain point, many of these same measures are sound investments in their own right, cutting energy bills and increasing a nation's macroeconomic efficiency and international competitiveness. If accompanied by a reduction in subsidies to the fossil fuel industry, the measures could also provide a level playing field for other, more benign energy sources and reduce a heavy burden on the national budget. Even if climate change turns out to be less severe than now feared, the insurance will pay for itself. And we have to consider the possibility that climate change will be much worse than the models predict. Remember the ozone hole.

The free-rider problem is a more serious obstacle to small bargains and unilateral national action, but it too may be surmountable. The key might be found in the Ministerial Declaration that came out of the Second World Climate Conference. As noted earlier, it welcomed the decisions by the nations of Europe, Japan, Canada, Australia, and New Zealand to stabilize emissions of CO_2 and other GHG emissions not controlled by the Montreal Protocol at around 1990 levels by the year 2000.[64] By agreement among themselves, these countries could probably eliminate the free-rider problem and open the door to less-than-global-scale bargains and national action. How? Simply by agreeing that they will not adhere to a global convention, once it is negotiated, unless that convention contains a clause recognizing the reductions in CO_2 emissions that national and regional actions achieve between a base year (say 1990) and the date on which the convention comes into force. Since this group of nations includes a number of large countries, it would have the political weight to make its agreement stick.

Although there are serious flaws in the arguments against unilateral action, the political forces that feed on them are real and no government can ignore them. Societal resistance must be expected to increase sharply as the costs of action go up and proposed policy changes have more drastic impacts on everyday life. Politicians will have to make "bargains" between electorates who want action and interest groups that resist it. This will not be easy. Not all socially and economically advantageous measures are adopted by governments. Not all gains for society will benefit every potential blocking coalition. Major energy-savings programs, for example, will affect powerful and politically sensitive factions in transportation, utilities, and processing industries. Similarly, a removal of the subsidies that promote deforestation could provoke the forest industry and union members who benefit from them.

Political bargains on the environment could become increasingly difficult and, judging from the growing debate in the United States, Canada, Japan, Australia, and parts of Europe, political leaders anticipate this and fear it.[65] The fall of the Netherlands' government in 1988 due to a series of proposed environmental initiatives, including the removal of a transportation subsidy, did not go unnoticed. History shows that, to succeed major policy initiatives in any field must be linked to measures to compensate losers and gather sufficient coalitions of political support. This is doubly true of measures to encourage more sustainable forms of development, since they often challenge deeply rooted codes of political behaviour. Take environmental taxes, for example. Even though they can be introduced in ways that do not increase the overall tax burden, as discussed in chapter 2, citizens may well be suspicious of a government's intentions. The political barriers may be surmountable if it can be made clear that the total tax burden will not increase, that lower-income groups will not suffer, that incomes and employment are not threatened, and that the new environmental focus of taxation serves a compelling social need.

Some countries have been more successful in this than others. In committing themselves to major reductions in CO_2 emissions, the nations of Western Europe and Japan clearly expect to succeed in securing effective bargains with competing national interests. Sweden, Finland, and other European countries are moving unilaterally to impose carbon taxes to limit fossil fuel emissions.[66] Germany is considering doing so where the policies involved can be justified for other reasons. The German government has also recognized that technologies developed now to combat global warming and acid rain could have an enormous market potential when other countries decide to act.

Global warming is not the first global environmental issue to come to the top of the political agenda, nor will it be the last. Consequently, it is important to develop fully the concept of bargains, and the funding arrangements to facilitate them. The long-term goal is an arrangement to promote environmental sustainability on a global scale and on a broad range of issues—forests, soils, species and ecosystems, atmospheric issues, ocean issues, hazardous materials, and fresh water. This will require a number of globally comprehensive conventions and protocols. They could be a long time coming and, in the meantime, it is important not to let the best be the enemy of the good. Rather, one must be opportunistic and actively promote deals with a wide range of actors, bilaterally and regionally as well as globally. The aggressive pursuit of smaller deals would build trust. They would also offer the opportunity to move around potential blocking coalitions that could obstruct more comprehensive deals; they would generate information on what works and what doesn't work; and they could take advantage of the progressive changes in environmental political values. A Grand Global Bargain could be the sum of 1,000 small bargains.

5

Challenges and Prospects for a Sustainable Future

ЛЛЛЛЛЛЛЛЛЛЛЛЛЛЛЛЛЛЛЛЛ

In June 1992 world leaders will assemble in Rio de Janeiro for the first Earth Summit on environment and development. They face an enormous challenge. They must initiate significant reform of the policies and institutions that are behind the steady depletion of our basic ecological capital, the accelerating degradation of the environment, and the growing threats to essential life-support systems. They must commence a process of basic change in the way we conduct domestic and international economic relations and in the way we make decisions in government, industry, and the home. They must begin to reshape our international institutions for an age of total interdependence.

If world leaders in June 1992 can start the necessary transition to more sustainable forms of development, the next two decades could see a gradual improvement in our capacity to manage interlocked economic and ecological interdependence. Instead of a future of progressive ecological collapse, economic decline, social strife, and conflict, they could set the stage for a secure and sustainable future for a majority of nations and peoples, with a steady improvement in the human prospect. The U.N. General Assembly has wisely invited heads of state or government to the Rio conference, giving the conference a unique capacity to make the fundamental decisions required and to set new directions for our common future.

The process of preparing for the Earth Summit is well under way. The Preparatory Committee is served by a dynamic secretary general,

Maurice Strong, and is open to all members of the United Nations. It is anticipated that the heads of state will be asked to deal with four key items. First, they will be asked to adopt an "Earth Charter" setting out broad directions for development and embodying new principles to govern relationships between governments, peoples, and the planet in the twenty-first century.[1] Second, they will be asked to adopt an agenda for action—"Agenda 21." As presently conceived, it will set out an internationally agreed-upon work program, including targets of national and international performance for each of the critical issues as discussed in chapters 1–4. The agenda will include the estimated costs of achieving those targets, and ways and means to provide the necessary financial resources. It will also contain an operational commitment to transfer the needed technologies to developing countries. Most importantly, it will designate the national and international agencies that will bear responsibility for the first phase of implementation, tentatively set for the last seven years of this century. Third, the summit will be asked to review the negotiations underway on several conventions, in particular those to limit global warming, halt net deforestation, and preserve the planet's biodiversity. Some participants have expressed hope that these conventions will be ready for signature. Fourth, the summit must initiate significant reform of our international institutions to enable nations to manage global interdependence and to implement Agenda 21 as well as the conventions.

This is a very tall order under the best of conditions, and conditions are far from the best. World opinion clearly wants the summit to succeed. A U.N. poll in 1990 found that public awareness and concern about the environment were at the highest levels ever recorded in the South as well as the North. This public pressure was obviously behind the decisions by Western European and some other governments to move ahead on a number of issues, including forestry and global warming, as discussed in chapter 4. But it had very little effect on some other key governments. There is scant evidence to suggest that these governments, led by the White House, have changed their historical resistance to any performance targets, or to the realities of technological transfer, or to any significant commitments on financing. They are not yet convinced that the policy changes required to reduce atmospheric concentrations of carbon dioxide or to conserve biological diversity or to transfer environmentally and economically competitive technologies to developing countries will be to their short-term commercial or political advantage. And they appear unwilling, if not unable, to think beyond the short term.

While public pressure has not yet forced some key Western governments to act, increasing awareness of threats to the environment has provided developing countries with leverage that they didn't have before. As discussed in chapter 3, they can trade their participation in new international environmental arrangements against Western agreement to address the related issues that concern them most—better trade access to Western markets, debt reduction, technology transfer, and additional aid and investment, including special funds to cover the net cost of their participation. This tactic proved quite successful in the negotiations to strengthen the Montreal Protocol on the ozone layer, and the prospects are that it will be used in the negotiations on the new conventions on forestry, conservation of species, and global warming. While this may be a sound and necessary tactic for developing countries faced with the intransigence of some key governments, its use could prolong the negotiations on certain issues.

A Herculean effort would be required to ensure a successful summit without the compelling distractions of conflict in the Middle East and political and economic instability in Central and Eastern Europe, including the Soviet Union. With them, the task seems formidable in the extreme. In early 1989, the world was celebrating the end of the cold war and looking forward to a "peace dividend," with military spending channeled into international development, the protection of the environment, and the building of a durable peace. In early 1990, this bonus was absorbed by a major conflict in the Persian Gulf and fear of possibly explosive change in the Soviet Union. The public's attention has been diverted by these events, and other equally compelling distractions could of course appear at any time.

It is hard to predict the impact of developments in the Gulf and the Soviet Union on basic concerns about the destruction of the environment. Experience since the late 1960s suggests that they may not be serious or of long duration. The 1970s and 1980s saw a steady rise in the underlying trend of environmental awareness and concern despite stagflation, recession, and smaller wars. Growing political and economic instability has brought a temporary halt to the ecological reconstruction of Eastern Europe and the Soviet Union but, in the longer run, this can only serve to raise concerns about the health, social, and economic consequences of continued ecological decline. The Gulf conflict has seen environmental destruction used as a weapon of war, and this could heighten the growing conviction that environmental risks pose the greatest threat to our common security. These and other events could well reinforce the pressure on leaders to make the Earth Summit a success.

It is easy to understand why the United Nations decided to hold a conference on the critical issues of environment and development in 1992. At no time in history has the world community faced as many important decisions as it does today. Whether these decisions are taken deliberately or by default, they could determine the fate of humankind. The prospects of these decisions being taken deliberately and positively are obviously better if they are preceded by a conference to build awareness and marshal the necessary political resources. This is perhaps even more true in 1991 than it was in 1988 when the General Assembly made its decision to hold an Earth Summit.

Great opportunities exist for a range of bargains to reduce the mounting threats to the environment and economic development, and to peace and security, as we have seen in chapters 1–4. Great obstacles also exist, and the instability in the Soviet Union and Eastern Europe and conflict in the Gulf simply add to these obstacles. "The earth is one but the world is not,"[2] and it is the world as it is that must confront these issues. We begin this final chapter with a discussion of the conventions that are being negotiated in preparation for the Earth Summit. We consider a number of options and, in the process, we again raise the question of whether significant bargains to address global issues should wait for the successful conclusion of global conventions, or whether a more opportunistic approach should be adopted, encouraging bargains at all levels and running with those that can be implemented now. We then consider the major tasks involved in negotiating and implementing many small bargains or a few comprehensive international conventions. We also explore the adequacy of existing institutions and the need for reform. We conclude with a look ahead.

International Conventions on Global Change: The Options

Three major international conventions are currently being negotiated— a world climate convention, a world forestry convention, and a convention to conserve biodiversity. In each case, the objective is to have "something" ready to sign at the Earth Summit in June 1992. The occasion could be an unparalleled global photo opportunity, hard for any president or prime minister to resist. Unless great restraint is shown, however, the form and content of the conventions presented to the heads of state could be determined by the pressure to create the photo opportunity rather than the issues themselves.

The diplomatic effort to launch a convention on climate change was given a major push at the June 1988 Toronto Conference on the Changing Atmosphere. Several models were explored and, for a short while after the conference, those wanting to negotiate an "umbrella" convention appeared to have the upper hand. In February 1989 the government of Canada convened a meeting in Ottawa to consider such a convention. It was called a "law of the atmosphere" and was designed to embrace air pollution, acid rain, ozone depletion, and global warming.[3] The term, of course, conjured up the famous Law of the Sea, which established the 200 mile economic zone and thereby put an additional 35 percent of the ocean's surface under national control for the management of natural resources; it also provided an international framework for the management of the resources of the seabed beyond national jurisdiction. While some experts felt that the Law of the Sea was an appropriate model for the atmosphere, most disagreed. They felt that the issues involved in the atmosphere were very different from those involved in the seas. Moreover, the current levels of use of the atmosphere are much higher than the oceans, they involve everybody directly, and vested interests such as the energy, utility, and transportation industries are much stronger than those involved with fisheries and exploring the seabed. As a result, the dissenters argued, the negotiations on a law of the atmosphere could be even more complex and time-consuming than those associated with the Law of the Sea. This was a dismal prospect, since the Law of the Sea took over 15 years to negotiate. Even though finally deposited for signature in December 1982 it still has not come into force, although many experts assert that large sections of it now have the force of recognized law. The Law of the Sea model was dropped following the Ottawa meeting, at least for the time being.

The dominant model today is the framework convention. It has been used on many occasions—long-range transport of air pollution, regional seas, ozone depletion—and is now being considered for the conventions on climate change, forestry, and biodiversity. There are two broad options for a framework convention. This first is an "empty" framework, which we will call Mark I. The second is a "substantive" framework, which we will call Mark II.[4]

A Mark I framework convention would be limited to a few noncontentious provisions. It might be limited, for example, to some basic definitions, some general principles concerning the responsibilities and rights of states, and some declarations of good intention concerning measures to prevent, reduce, or control climate change. If it were daring, it might also include certain obligations to cooperate in consultations and in monitoring, research, the development and transfer of technology, and the transmission and exchange of information. It could also provide for a broad coordinating structure and, probably, a secretariat. But it would include no targets and, hence, have no need for agreement on measures to achieve them.

The experience with the 1985 Vienna Convention on the Protection of the Ozone Layer is often advanced to support the "empty" framework convention approach. It made provision for subsequent protocols that might involve specific targets, and this did in fact happen in the Montreal Protocol only two years later—a remarkably short period of time.[5] But the experience with the 1979 ECE Convention on Long-Range Transport of Air Pollution was quite different. In spite of the evidence of enormous economic damage from acidification, industrialized countries have only in the last few years begun to arrive at agreements on reduction targets and policy measures and schedules to reach those targets. It is perhaps needless to add that climate change is an infinitely more complex issue than either the ozone layer or acidification of the environment.

The most important role of an "empty" Mark I framework convention is, as its name implies, to provide a framework for later protocols on the hard issues, protocols that could contain specific targets and measures to achieve them. Given the complexity and contentiousness of these issues, some feel that an empty framework may be all that is possible within the foreseeable future. And, they argue, even a symbolic agreement may put in place new constituencies such as an international research project or a secretariat working on the issues, and may build momentum for a more substantive agreement in the future.

Others are more skeptical. The fact is that empty framework conventions are politically very attractive. First, they are easy to negotiate. After all, they are little more than a wish list. Signing an empty framework convention is like an author signing a contract with a publisher on the basis of a table of contents and a statement of good intentions to have the text written (perhaps by somebody else) at some later date. Second, they allow national leaders to gain enormous political credit at no political cost. This is important in 1992. A number of leaders will face national elections just before or shortly after the Brazil conference and, if the environment continues to ride high in the polls in their countries, their record on environment could be a major issue. The signing ceremonies, with Rio as a backdrop, will provide a wonderful set of photo opportunities and would enable leaders to return home having appeared to make good on their environmental pretensions.

The great danger with a Mark I framework convention is that it will deceive the public into believing that some significant progress has been achieved. The public doesn't know the difference between an "empty" convention and a "substantive" one. It could easily be deceived into believing that something significant had been done. The heat would be off governments to act before any meaningful agreement is reached on concrete measures: goals, staged targets to achieve those goals, new money and strong policy interventions to achieve the targets, and effective systems of monitoring and enforcement. With public pressure relieved, any further agreement would be unlikely. The most striking example of this was the partial nuclear test-ban treaty signed by the United States, the Soviet Union, and the United Kingdom in August 1963, which effectively defused political momentum for a comprehensive ban.

An alternative, which the authors would support, is to take the time needed to negotiate a "substantive" framework convention—Mark II. A Mark II convention would contain a clear commitment to the goal of stabilizing GHG concentrations in the atmosphere and a set of phased targets to achieve that goal. The protocols, to be negotiated in parallel with the convention, would pick the agreed-upon targets and set out the designated means to achieve them, including any new financial arrangements.[6] A Mark II convention would also lay out the foundations for a simple but effective mechanism to provide leadership, overall political direction, and broad coordination of the work to be done. At a minimum, a Mark II convention should not be signed unless and until at least a protocol on energy emissions and one on forestry can be signed, along with measures for financing. Later protocols could provide for appropriate measures concerning the hitherto uncharted territory of methane and nitrous-oxide emissions.

Can a Mark II framework convention for climate change, forestry, and biodiversity be negotiated in time for the June 1992 Earth Summit in Rio? It all depends on the will and actions of world leaders, driven as they are by events and public opinion. At the moment, however, it seems very unlikely. If negotiations through 1992 do not produce a substantive, Mark II framework convention with goals, targets, and a few key action protocols that can be signed at the same time, it would be better in our view to opt for no convention than to settle for an empty Mark I framework. A political "success" purchased at the price of an empty framework would reduce pressure for real action, perhaps for years to come. The chastening experience of a political "failure," on the other hand, could sharpen the pressure and maintain the momentum of the negotiations until real success is achieved. The worst thing that the Earth Summit could do is to let governments "off the hook." Rather, it should increase societal pressure on governments and "make them sweat it out" until such pressure has been translated into a political decision to act. Failing a Mark II convention, the Earth Summit could become an occasion for a serious debate on the hard issues, providing leadership and direction to the ongoing negotiating process.

Seizing Opportunities Now

The crucial question at this stage is whether significant bargains to address global issues need to await the successful completion of negotiations on a series of global conventions, or whether it might not be wiser to adopt a more opportunistic approach, encouraging actions at all levels and going with those that can be undertaken now: unilateral actions; bilateral actions; regional actions—actions supported by small groups of like-minded nations or by entire regions. Even though it may seem messy and theoretically less efficient, this pluralistic "let a thousand flowers bloom" approach offers the major benefit of having a realistic chance of some early successes. It would enable us to move in smaller steps with smaller institutions, to move more efficiently and effectively, and to get results more quickly. It would also provide us with an invaluable learning experience. It would teach us the dos and the don'ts of, for instance, large-scale afforestation projects, and would be a source of experience and credibility for other countries. It would develop markets and new knowledge and technology. It would also set examples and exert pressure on potential "free riders," thus furthering the process of international policy development. Above all, it would help nations gain understanding about shaping global bargains that answer the needs of all sides and that link issues of environment with those of development, trade, economic policy, and security.

The overriding goal is to promote environmental and economic sustainability on a global scale and on a broad range of issues: forests, soils, species, chemicals, the atmosphere, oceans, and fresh water. This eventually will require a number of globally comprehensive conventions and protocols. But they could be a long time coming and, in the meantime, it is important not to let the best be the enemy of the good. In this divided world, with its skewed power structure, it is often wiser to be opportunistic and actively promote deals that can be concluded now with a wide range of actors, bilaterally and regionally as well as globally.

Institutional development and reform is a gradual evolutionary process. Rather than going for a home run in the form of a series of Grand Global Bargains, we should aim for a run of singles—building on the current institutions and modifying their charters and their operations as we learn from our experiences. The key to action and to winning is to get on base and to play the game as it develops. The aggressive pursuit of a series of smaller bargains would build trust. This course would also offer the opportunity to move around potential blocking coalitions that could obstruct more comprehensive deals; it would generate information on what works and what doesn't work; and it could take advantage of the progressive changes in environmental values and domestic political pressure. A Grand Global Bargain could be the sum of 1,000 small bargains.

Institutions for Small Bargains

New or expanded tasks will need to be performed for each bargain negotiated. They could range from providing the necessary forums for discussions and negotiations to marshaling, managing, and distributing the funds committed to the bargain. Depending on the nature of the bargain, they could include procedures for sharing information, coordinating research and development, facilitating negotiations on the transfer of specific technologies needed to implement the bargain, and

FIGURE 5.1 Some essential tasks

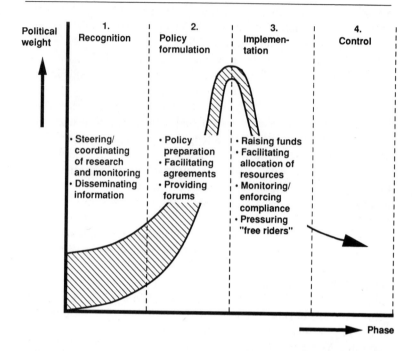

monitoring implementation, enforcement, and compliance. They could also include coordinating efforts by other international agencies party to the bargain, including those responsible for aid, trade, and security. Figure 5.1 provides an indicative list of some key tasks that will need to be performed, linking them to the policy life cycle discussed earlier (see Figures 3.2 and 3.3). As we proceed from the recognition phase of a problem to the policy formulation, implementation, and control phases, the tasks (and the institutional support needed) shift from research and monitoring to raising and allocating funds and to pressuring "free riders" to join the club and pay their dues. Marshaling know-how, in the form of people, documentation, and technology, will also be a vital task in many bargains, and this will require substantially increased funding for scientific research, especially in the Third World.

In an ideal world, potential partners in a bargain would agree about what needed to be done before naming the institution or institutions to do it. In real life, these matters are usually considered together. In fact, the opportunity to expand institutional turf (and improve competitive position) can be a powerful motivating force behind a potential bargain. Governments, as a rule, will and should rely on existing organizations, including networks and "switchboard" or "clearinghouse" mechanisms. New agencies, including new supercoordinating mechanisms, may be needed at some stage but should be considered only after other options have been exhausted.

Fortunately, few small bargains should have to await the development of new institutions. A rich variety of centers exist, inside and outside of government, that can initiate, strike, and "guarantee" bargains. In addition to governmental and intergovernmental agencies, they include private foundations and institutes of science, engineering, and public policy. Small bargains involving just a few countries can be negotiated through existing national and bilateral institutions, often perhaps by foreign-aid and environment ministries supported by science, industry, trade, export–credit, and other institutions. A range of organizations also exists that can facilitate larger bargains at the regional level. The West is particularly well-served by organizations like the OECD and the International Energy Agency, as is Europe by the Commission of the European Communities (CEC). The U.N. Economic Commission for Europe (ECE) has played a significant role in bridging East and West on a range of issues, including environmental protection and resource management, as has the International Institute for Applied Systems Analysis (IIASA). The South is not nearly as well-served by effective regional bodies but a number are beginning to evolve, such as the Southern Africa Development Coordination Conference (SADCC), the Organization of American States, the Caribbean Community (CARICOM), the South Asian Association for Regional Cooperation (SAARC), and the Association of South East Asian Nations (ASEAN). The regional development banks must also be mentioned here. Some of them, most recently the Inter-American Development Bank (IADB), but also the Asian and African Development Banks, have begun to adopt proactive stances on issues of global change.

At the global level, bargains can be facilitated by an increasingly developed family of United Nations organizations. Prominent among these are the United Nations Environment Program (UNEP), the World Meteorological Organization (WMO), the World Health Organization (WHO), the Food and Agriculture Organization (FAO), and the United Nations Educational, Scientific and Cultural Organization (UNESCO).

Organizations such as the United Nations Conference for Trade and Development and the General Agreement on Tariffs and Trade will have increasingly important roles to play. Just to mention their names is to suggest a number of priority areas in which bargains are urgently needed. An increasingly important role in the areas of funding, investment finance, and policy reform must be played by the United Nations Development Program, the World Bank, and the International Monetary Fund (IMF).

Outside of the "official" system, we find a large number of organizations, some with an enviable track record on environmental issues. They include many national, regional, and international scientific academies and bodies such as the International Council of Scientific Unions, and national and international nongovernmental organizations such as the International Union for the Conservation of Nature (IUCN). And then there are the philanthropic institutions—exemplified by the great private foundations—and the universities, which are among the world's most vital engines of democracy, pluralism, and change. These groups and hundreds of others play an indispensable role in designing and implementing measures to deal with the issues of global change. Their capabilities, focused on specific bargains, will be even more important in the future.

Strengthening Institutional Capacity

Clearly, we do not start empty-handed. An extensive institutional structure exists internationally to support the negotiation and implementation of global bargains. But, as repeated studies have confirmed, this structure lacks political strength, cohesion, depth, and financial capacity. Many of these studies have focused on the U.N. system and on ways and means to strengthen it.[7] Most recently, the U.N. Preparatory Committee has taken up these studies and it is expected that a number of recommendations for reform will be brought before the Earth Summit in Rio in June 1992. We will not attempt to anticipate their recommendations, but will limit ourselves to a few comments about certain key agencies with responsibilities for environment, development funding, and overall political direction.

The startling thing about many U.N. agencies concerned with the environment—for example, UNEP and WMO—is that they have been able to accomplish so much with so little.[8] UNEP led the work on the Vienna Convention and the Montreal Protocol on the Ozone Layer, and has been a key participant in the search for solutions to critical environmental problems such as desertification and marine pollution. Both UNEP and WMO played leading roles in support of the work leading up to the Second World Climate Conference in November 1990. WMO operates the Global Ozone Observing System and it led the major scientific assessments of depletion of the ozone layer in cooperation with NASA and UNEP. These high-profile activities, however, mask the essential weakness of the organizations. The fact that they lack political strength (their "clients" being environmental and meteorological agencies that, in most governments, are at the bottom of the political pecking order) and that their budgets are derisory in relation to their mandates and the huge tasks they have been asked to perform (less than $50 million a year for UNEP, and about $23 million for WMO) has taken its toll. As noted, the Plan of Action to Combat Desertification (PACD), endorsed with great fanfare at two international conferences, is paralyzed for want of the promised funding.[9] So, in fact, are many other plans and agreements, including the much-heralded regional seas agreements on the Mediterranean (i.e., the "Blue Plan" signed by all the coastal nations), the Caribbean, and eight other seas involving a total of 120 countries. Each one of these regional seas agreements included a table of contents for later protocols that were to set out concrete action programs and provide the funds needed to implement them. Leaders of the day amassed enormous political capital at signing ceremonies, during which they bound their successors in office. So far, unfortunately, their successors in office have refused to be so bound to make the hard decisions on priorities, programs, and funding.

UNEP plays a central role among the United Nations and other international agencies dealing with the environment. Several studies have called upon governments to strengthen its political and financial capacity, to provide it with a stronger mandate in the area of environmental protection, and to extend and enhance its status within the U.N. system.[10] Some would prefer that it become a specialized agency with a much larger fund at its disposal; others that it be strengthened as a part of the central structure of the United Nations, with a stronger coordinating and policy role and with development funding centralized in the

UNDP. So far all recommendations have fallen on barren ground, but there is some hope that governments will act on at least some of them at the Rio summit.

Several studies have also called upon governments to strengthen the environmental direction, sensitivity, and capacity of the World Bank, the IMF, and the regional development banks. The World Bank, the largest single source of development lending, is especially important because it exercises significant policy leadership on other donors and on developing countries. It has been and remains under attack for the lack of environmental sensitivity in its lending practices. Since 1987, however, the Bank has made some major changes in its structure, staffing, and processes to give environmental concerns a higher place in its investment lending decisions. In 1990 it established a Global Environmental Facility based on voluntary contributions of $1 billion to fund projects in a variety of fields. These changes are encouraging, but the experience to date strongly suggests that they will not be fully effective unless they are accompanied by a fundamental commitment to sustainable development and corresponding changes in the bank's institutional culture. With such a commitment, the World Bank could perhaps grow into the role of serving as the international clearinghouse. The IMF also exerts a major influence on policy and economic practices in developing countries, in particular through the conditions it attaches to its reform packages. Unlike the development banks, the IMF has shown no signs as yet of making the changes necessary to reflect environmental concerns in its decisions.

Reforming Our Great International Institutions

The massive changes occurring in the relationships between the world of nation states and the earth and its biosphere have not been accompanied by corresponding changes in our international institutions. The Earth Summit must be a driving force for a major reform of our key institutions, from the United Nations Security Council to the U.N.'s specialized agencies, and from the IMF, the multilateral banks, and GATT to national ministries of foreign affairs, finance, and economic planning. The World Commission on Environment and Development put forward a large number of recommendations for such reform, and new proposals have since been added to the list. Some, as noted above, take an incremental approach, building on existing institutions. Other recommendations foresee more fundamental change involving some pooling of national sovereignty.

The Hague Declaration of March 1989, for example, calls for a new international authority with responsibility to prevent further global warming.[11] It recognizes that this can be done either by establishing a new institution or by extending an existing one. But—and this is the most significant element in the declaration—it accepts that the institution should be able to make certain decisions by means other than consensus. And it anticipates measures to enforce compliance, with appeals against such measures being placed before the International Court of Justice. More than 30 nations have signed the Hague Declaration. Although it is not binding, it should help the Rio Summit break out of traditional mindsets in judging other proposals for institutional change, something that is clearly needed. There have been few indications that major nations are prepared to relinquish any, let alone a substantial, part of their sovereignty to an international body. Yet, these same nations have accepted a steady encroachment on their sovereignty by the forces of economic interdependence. This erosion will accelerate as a result of ecological interdependence and its rapid meshing with economic inter-dependence. In the longer term, therefore, conflict between national sovereignty and the authority required by international institutions to address critical issues may become less intense.

International legal regimes have also been rapidly outdistanced by the accelerating pace and scale of change. New norms for state and interstate behavior are urgently needed. At a minimum, the world community requires some basic norms for prior notification, consultation, and agreement with neighboring states on activities likely to affect them. States must also accept the obligation to alert their neighbors in the event of an accident (e.g., Basel, Chernobyl, Exxon *Valdez*), including the obligation to provide compensation for any damage done. Similar norms and obligations must gradually evolve for protecting the global commons and future generations. The proposed Earth Charter will provide the summit with an opportunity to push the international system in this direction.

Proposals for reform of the U.N. system go well beyond UNEP, the specialized agencies, and the multilateral development banks.[12] Some of the most important options being considered concern the system's capacity to provide world leadership, overall political direction, and essential coordination on global issues of environment, development, peace, and security. It has been proposed, for example, that the Security Council should periodically devote a special session to environmental threats to peace and security. The Soviet Union has hinted that it might support a new Environment Council, equal in authority to the Security Council, but perhaps without the right to veto. The Trusteeship Council is coming to the end of its mandate and the secretary general of the Earth Summit, Maurice Strong, has suggested that this body be transformed into a forum in which the nations of the world would exercise their "trusteeship" for the integrity of the planet as a whole, including the global commons. The idea of a new "Earth Council," another one of these innovative formulations, reflects the quality of imagination and the level of ambition that needs to be applied in developing new forms of governance to guide the planet through the next turbulent decades.

Questions of sovereignty are fortunately not involved in all proposals for useful change. It has been suggested, for example, that UNEP should be given the resources necessary to act on behalf of the U.N. secretary general as an environmental "secretariat" for the Security Council or a revitalized Trusteeship Council. It would undertake the monitoring and provide the analyses and assessment reports needed for special sessions of the council to deal with environmental threats to global security. Other proposals would strengthen UNEP's mandate and resources in the area of environmental protection, but they would establish a new, standing U.N. Commission on the Environment, or a Special Independent Commission on Environment and Sustainable Development to service the Security or Trusteeship Councils through the secretary general.

Still other proposals would underpin these initiatives with a continuing forum on environment and development, a political assembly representative of all nations to provide leadership, guidance, and support to the U.N. system as a whole in its work in these areas. Such a forum might be created within the framework of a Mark II convention in order to provide leadership, overall political direction, and broad coordination of work under the convention. Called, perhaps, a World Environment and Development Forum (WEDF—Figure 5.2), it would bring together ministers representing the signatories to the convention at least once a year to evaluate progress and decide on necessary actions. Since any successful global bargain on climate change would have to address a range of economic, trade, technology transfer, and equity issues, nations may wish to send not only ministers of environment and resources, but also senior ministers of economic agencies, development planning departments, or even foreign affairs.

Experience within OECD, the European Communities, and the ECE has shown that regular meetings of ministers can be a powerful force in consensus building and eventual decision-making. The WEDF would be the highest platform for policy development and coordination on issues of climate change. As consensus grows, it might develop into an "Earth Council" as envisaged at the Hague Conference, with much greater authority to make decisions and to enforce regulations. Or it could become a standing commission on environment and sustainable development within the U.N. framework, as discussed above.

The WEDF would need to be supported by a small permanent bureau. A group of senior officials would be essential to prepare policy recommendations for its consideration and to set up channels for the implementation of its decisions. UNEP and the WMO might perform these roles. The Intergovernmental Panel on Climate Change, which involves senior officials from participating governments, could undertake scientific assessments and analysis of policy options on a continuing basis. In addition, the WEDF could play a major role in strengthening the information, communications, policy development, and compliance-monitoring "infrastructure" on global warming. This would require: (1) an improved fact base; (2) a program to build public awareness about climate change and to educate professionals in possible responses; (3) a program to further develop policy tools, including those that are necessary for assessment of the costs and benefits of inaction and the socioeconomic impacts of response strategies; and (4) significant emphasis on monitoring national compliance with greenhouse gas reduction targets and other commitments. Given the high stakes involved in global warming, the costs for this work would be relatively minor—probably in the order of $50 million to $100 million per year.[13] The establishment of a World Environment and Development Forum as a meeting place for world leaders would provide a first significant step toward recognizing our global interdependence and the interlocking of the world's economy and the earth's ecology. It would provide the "let a thousand flowers bloom" approach with a steering mechanism.

The menu of options can be expected to grow rapidly as we move into the negotiations on a climate convention and the preparations for the Earth Summit in June 1992. Some options mentioned earlier relating

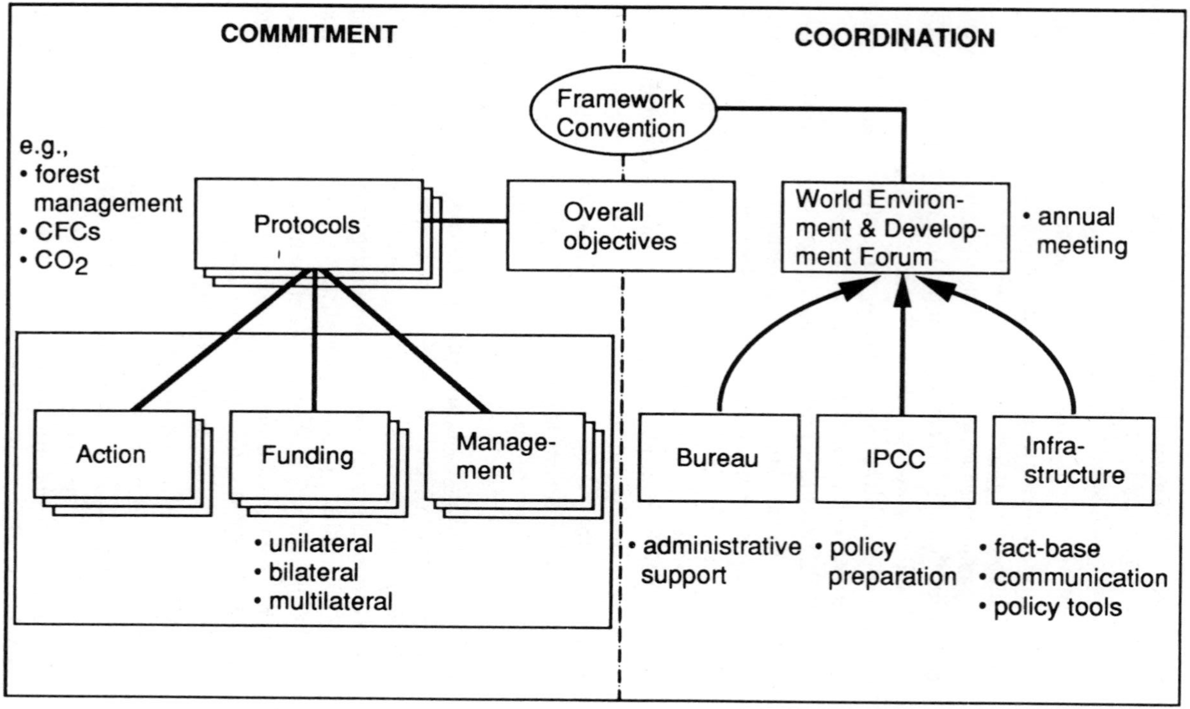

FIGURE 5.2 Framework for international coordinating mechanism

to the national and regional levels could be put into effect rather quickly, especially if they took the form of bargains on institutional development and training. Others relating to the U.N. system could be put into effect by the General Assembly, while still others would require amendments to the U.N. Charter.

The Earth Summit: Rio de Janeiro, June 1992

The debate has just begun on the policy and institutional changes needed to reverse environmental degradation and move toward more sustainable forms of development. What happens or fails to happen on the Road to Rio will be crucial. The negotiations leading up to Rio will force governments to decide—and leaders to reveal—the extent to which they are really prepared to go beyond rhetoric and make the difficult decisions that are now needed. Western leaders must demonstrate enlightened leadership. Only they command the economic resources, the technologies, and the political resilience to accommodate significant change. Only they can initiate the restructuring of international economic and political relations needed to reverse the tragic flow of capital from the poorer to the richer countries and ensure that developing countries get equitable access to the technologies needed to support sustainable development. A breakthrough in these two areas will be the key to the success of the Earth Summit. Failure would be an enormous setback to North–South relations and would likely cripple prospects for a new global alliance to secure the future of our planet.

Many of the leaders who gather in 2012 to mark the fortieth anniversary of the Stockholm Conference and the twentieth of Rio will have been youngsters at the time of those events.[14] How will they assess what this generation has done and not done? How harshly will they judge those who made or avoided the crucial decisions during the previous decades?

Their assessments and judgments will depend on what they see around them. Will they find less or more poverty and hunger, less or more environmental pollution and degradation, less or more economic and social justice within and among nations? Will they have inherited countries still truculently stuck in obsolete notions of national sovereignty or countries that have embraced their common interests and dedicated themselves to working together to give renewed meaning to "We the peoples of the United Nations''? Will they lead countries that have found pathways to sustainable development or countries still locked in a downward spiral of ecological and economic decline?

The answers to these questions will be determined largely by what happens or fails to happen before and at Rio in June 1992. At best, Rio could serve to break the inertia by which national policies and institutions continue to reinforce downward trends. It could place the world community on a new course toward a sustainable future. If Rio is successful, the year 2012 should see population growth under control, the development needed to meet human needs and aspirations achieved on a sustainable basis with greater equity within and between nations, systems of public and private economic incentives in place to promote greater energy and resource efficiency, the development of renewables and substitutes for fossil fuels, the development of low and nonwaste technologies, full recycling, and cradle-to-grave management of all hazardous materials. Emissions of greenhouse gases and ozone-depleting substances should have been reduced to safe levels, the planet's biodiversity and genetic resources protected, and net deforestation reversed. The Four Horsemen will no doubt continue to thrive but they should be walking rather than galloping through a much smaller part of the world than they are today. Most of all, 2012 should see a new global partnership expressed in a revitalized international system in which an Earth Council, perhaps the Security Council with a broader mandate, maintains the interlocked environmental and economic security of the planet.

The Earth Summit will likely be the last chance for the world, in this century at least, to seriously address and arrest the accelerating environmental threats to economic development, national security, and human survival. It will certainly be the last major chance for the present generation of leaders and decision-makers to fulfill their basic obligations to their peers, today's youth, and future generations.

Appendixes

ЛЛЛЛЛЛЛЛЛЛЛЛЛЛЛЛЛЛЛЛЛЛЛ

1. Growth, Distribution, and Poverty

How quickly can a developing country expect to eliminate *absolute* **poverty?** The answer will vary from country to country, but much can be learned from a typical case.

Consider a nation in which half the population lives below the poverty line, and where the distribution of household incomes is as follows: The top one-fifth of households have 50 percent of total income, the next fifth have 20 percent, the next fifth 14 percent, the next fifth 9 percent, and the bottom fifth just 7 percent. This is a fair representation of the situation in many low-income developing countries.

The income of the bottom fifth would have to double (to the initial level of the third quintile) to bring the fraction in poverty down from 50 percent to 10 percent. Consider two cases, one in which 25 percent of the incremental income of the richest one-fifth is redistributed equally to the others, and one in which there is no redistribution of increases in income. The number of years required to double the income of the bottom fifth in the two cases would be:

- 18 or 24 years, if per capita income grows at 3 percent,
- 26 or 36 years, if it grows at 2 percent,
- 51 or 70 years, if it grows only at 1 percent.

So, with per capita income growing at only 1 percent a year, the time required to eliminate absolute poverty would stretch well into the next century. If the aim is to ensure that the world is well on its way to sustainable development during the first part of the next century, it is necessary to aim at two things: a minimum of 3 percent per capita national income growth *and* vigorous policies to achieve greater equity within developing countries.

SOURCE: World Commission on Environment and Development, *Our Common Future* (New York: Oxford University Press, 1987), p. 50.

2. Strategic Imperatives for Sustainable Development

1. Growth sufficient to meet human needs and aspirations.
2. Policies to increase equity within nations and between developed and developing countries.
3. Policies to reduce high rates of population growth.
4. Policies to conserve and enhance the resource base.
5. Policies to ensure a rapid reduction in the energy and resource content of growth.
6. Institutional change to integrate environment in economic decision-making.

SOURCE: Adapted from *Our Common Future*

3. Standard "First Generation" Agenda (Indicative)

Pollution Issues

Air pollution
Acid rain
Climate change
Water pollution
Chemicals
Hazardous wastes
Nuclear wastes

Natural Resource Issues

Deforestation
Loss of genetic resources
Loss of cropland
Soil erosion/desertification
Water resource management
Depletion of marine resources
Parks, wetlands, and recreation
    areas

Urban Issues

Land use and tenure
Shelter
Water supply and sanitation
Social welfare, health, and
    education
Megacity
Managing urban growth

Management Questions

Monitoring and reporting
Investment analysis
Benefit–cost
    Cost-effectiveness
    Risk analysis

4. Sustainable Development Agenda (Indicative)

Economic development and environment
Population, human resources, and education
Urbanization and urban development
Energy, development, and environment[1]
Agriculture, forestry, development, and environment[2]
Industry, development, and environment[3]
International economic relations[4]
Governance and decision-making[5]
International cooperation

    Global commons
    Peace, security, development, and environment
    Financing development

[1]Includes air pollution, acid rain, climatic change, fuelwood, renewables, and alternative sources, from the "standard agenda."

[2]Includes soil erosion, desertification, loss of cropland, tropical forests, and biological diversity.

[3]Includes industrial safety, chemicals, waste management and resource recycling and recovery, and related issues.

[4]Includes policies concerning trade, development assistance, transnational corporations, and international externalities.

[5]Includes integration of environment and economy in national and international institutions of governance.

5. The Policy Life Cycle

Environmental issues vary widely in terms of their scope, sources, effects, risks, and social and economic consequences. A specific solution must be found for each separate issue. But most policy issues pass through a policy life cycle consisting of four separate phases with highly different characteristics determined by a critical policy-making parameter: the political weight attached to a specific issue at a given moment in time (see Figure 3.2).

Phase 1: Recognition of the problem The first question is: Do we have a problem? Signals that a problem might exist typically come from researchers and/or environmentalists, although there usually are initial differences of opinion regarding the nature and extent of the problem and its causes and effects. Thus, a key factor for success is the management of uncertainty. The demand for a suitable policy comes when the responsible government authorities, often as the result of a major incident—e.g., Seveso (Italy), Bhopal (India), Three Mile Island (Pennsylvania), Chernobyl (U.S.S.R.), Basel (Switzerland), Love Canal (New York), Exxon *Valdez* (Alaska), reach the conclusion that the problem must be solved.

Phase 2: Policy formulation Political discussion about the most appropriate measures and the correct distribution of costs is often fierce during this phase. As a result, press coverage is intensive and the key to success is often crisis management, with policy-makers emphasizing effectiveness rather than efficiency: whatever the cost, find a solution that works and get it through the legislature.

Phase 3: Implementation The actual implementation of the policy is often costly and has a major microeconomic impact on industry, power generation, transportation, agriculture, and/or private households. However, once it has been determined how the problem will be solved, the political and societal attention paid to it tends to taper off. With the shift toward operational management, more emphasis is placed on enforcing and particularly on streamlining regulations and procedures (re-regulation). Environmental policy-makers turn their attention to the efficiency of handling the problem. Decentralization of responsibilities to lower levels of government (state, municipalities) frequently plays a crucial role in this respect.

Phase 4: Control The fourth and final phase of the policy life cycle begins when the intended improvement in environmental quality has been realized and the problem is reduced to proportions that are considered politically and technically acceptable based on existing information. Environmental policy-makers now must ensure that the problem remains under control. As the policy is internalized throughout society, regulations concerning the issue can often be simplified and sometimes even abolished (deregulation). As a rule, vigilance remains necessary, however.

Although national priorities tend to differ and environmental policy might be more advanced in specific regions of the world, a rough overview presents an indication of the current state of development of major environmental issues in industrialized countries (see Figure 3.3).It should be noted that the situation may be widely different in different parts of the world. Issues that within some western countries may be in phase 4, in others may be in phase 3, and in developing countries may be in phase 2 or even phase 1. Nations are, as it were, out of phase and therein lie the opportunities for bargains.

SOURCE: Adapted from *Guests in Our Own Home*, 1986, Pieter Winsemius (McKinsey & Company, 1990).

Notes

```
ⅬⅬⅬⅬⅬⅬⅬⅬⅬⅬⅬⅬⅬⅬⅬⅬⅬⅬⅬⅬⅬⅬⅬⅬⅬⅬⅬⅬ
```

Chapter 1

1. Tom d'Aquino, "Till Death Do Us Part," Address to National and Provincial Round Tables, Winnipeg, April 1989. Ottawa: Business Council on National Issues, 1989.

2. The World Commission on Environment and Development, *Our Common Future*. Oxford: Oxford University Press, 1987. See also Jim MacNeill, "Strategies for Sustainable Economic Development," *Scientific American* 261 (September, 1989): 154.

3. Ruth Leger Sivard, *World Military and Social Expenditures, 1988–89*. Washington, D.C.: World Priorities Institute, 1989.

4. The World Bank, *World Development Report, 1989*. Washington, D.C.: World Bank.

5. Peter M. Vitousek et al., "Human Appropriation of the Products of Photosynthesis," *BioScience* 36, (June 1986): 368. Human transformation and appropriation is 40 percent of net primary productivity of all terrestrial ecosystems.

6. Thomas E. Graedel and Paul J. Crutzen, "The Changing Atmosphere." *Scientific American* 261, (September 1989): 58.

7. Bertil Bolin et al., *The Global Carbon Cycle, SCOPE 13*. Chichester, England: John Wiley, 1979.

8. The World Commission on Environment and Development, *Mandate for Change* Geneva: WCED, August 1984.

9. Norman Meyers, *Deforestation Rates in Tropical Forests and their Climatic Implications*. London: Friends of the Earth, 1989.

10. U.N. Economic Commission for Europe, *Forest Damage and Air Pollution: Report of the 1988 Forest Damage Survey in Europe*. UNEP, Global Environment Monitoring System, 1989.

11. L. Hallbacken and C. O. Tamm, "Changes in Soil Acidity from 1927 to 1982–4 in a Forest Area of Southwest Sweden." *Scandinavian Journal of Forest Research*, no. 1, 1986.

12. "Neuartige Waldschaden in der Bundesrepublik Deutschland," Das Bundesministerium für Ernährung, Landwirtschaft und Forsten, 1983; "Waldschaden Sernebungen," Das Bundesministerium für Ehrnährung, Landwirtschaft und Forsten, 1985.

13. Lester R. Brown, "Reexamining the World Food Prospect," in *State of the World 1989*. Washington, D.C.: WorldWatch Institute.

14. Jodi L. Jacobson, "Abandoning Homelands," in *State of the World 1989*. Washington, D.C.: WorldWatch Institute.

15. National Aeronautics and Space Administration (NASA), *Executive Summary of the Ozone Trends Panel*. Washington, D.C., March 15, 1988.

16. J. T. Houghton, G. J. Jenkens and J. J. Ephraums, eds., Intergovernmental Panel on Climate Change, *Climate Change, the IPCC Scientific Assessment*. Report prepared for IPCC by Working Group I, World Meteorological Organization and United Nations Environment Program. Cambridge: University Press, 1990.

17. Jill Jaeger, "Developing Policies for Responding to Climatic Change." A summary of the Discussions and recommendations of the workshop held in Villach (September 28–October 2, 1987) and Bellagio (November 9–13, 1987) under the auspices of the Beijer Institute, Stockholm. World Meteorological Organization and United Nations Environment Programme, April 1988.

18. *The Montreal Protocol on Substances that Deplete the Ozone Layer, 1987*, UNEP/WMO, 1987. Nations agreed on a 50 percent reduction in production and consumption of the major ozone-destroying CFCs by developed countries by 1992, with a 10-year time lag for the developing countries.

19. Jaeger, "Developing Policies for Responding to Climatic Change."

20. Richard Houghton and George Woodwell, "Global Climate Change." *Scientific American* 26 (April 1989): 36–44.

21. Report by a Commonwealth Group of Experts, *Climate Change: Meeting the Challenge*. Commonwealth Secretariat, Marlborough House, Pall Mall, London SW1Y 5HX, September 1989.

22. Environment Canada, *The Changing Atmosphere: Implications for Global Security: Toronto, Canada, June 27–30, 1988*. Conference statement, Environment Canada, 1988.

23. Eric Arrhenius and Thomas Waltz, "The Greenhouse Effect: Implications for Economic Development." Discussion paper no. 78. Washington, D.C.: World Bank, 1990.

24. Intergovernmental Panel on Climate Change, *Scientific Assessment of Climate Change*. Report to IPCC from Working Group 1, third draft. Bracknell, U.K.: Meteorological Office, May 2, 1990.

25. *Our Common Future*.

26. For a recent summary and update of the work of the WCED (1987), see Robert Repetto (1986), R. Kerry Turner (1988), David Pearce, Edward Barbier, and Anil Markandya, *Sustainable Development: Economics and Environment in the Third World*. London Environmental Economics Center: Edward Elgar Publishing, February 1989.

27. World Bank, "World Debt Tables 1989–90, External Debt of Developing Countries," vol. 1: "Analysis and Summary Tables," p. 78. Washington, D.C." World Bank, 1989.

28. Ingo Walter and J. H. Loudon, "Environmental Costs and the Patterns of North–South Trade." World Commission on Environment and Development, 1986.

29. Nafis Sadik, "Investing in Women." Helsinki, Finland: statement to United Nations Population Fund, May 12, 1989.

30. Robert Repetto, *Public Policies and the Misuse of Forest Resources*. New York: Cambridge University Press, 1988.

31. For a full discussion of ecologically perverse agricultural policies and their reform, see *Food 2000, Global Policies for Sustainable Agriculture*, the report of the Advisory Panel on Food Security, Agriculture, Forestry and Environment to the the World Commission on Environment and Development. London: Zed Books, 1987. See also chapter 5 of *Our Common Future*, and *World Development Report, 1986*, Part II: "Trade and Pricing Policies in World Agriculture."

32. Mark Kosmo, *Money to Burn? The High Cost of Energy Subsidies*. Washington, D.C.: World Resources Institute, 1987.

33. Ian Brown, "Energy Subsidies in the United States," *Energy Pricing: Regulation, Subsidies and Distortion*. Surrey Energy Economics Centre Discussion Paper No. 38. Guildford, England: University of Surrey, 1989.

34. Udo E. Simonis et al., "Structural Change and Environmental Policy, Empirical Evidence on Thirty-One Countries in East and West." Berlin: Science Center, July 1988. See also "Ecological Modernization of Industrial Society—Three Strategic Elements." In Franco Archibugi and P. Nijkamp, eds. *Economy and Ecology: Towards Sustainable Development*. Dordrecht, Boston, London: Kluwer Academic, 1989.

35. *World Development Report, 1989*.

36. Ibid.; and Eric D. Larson, Marc H. Ross, and Robert H. Williams, "Beyond the Era of Materials," *Scientific American* 254 (June 1986): 34–41.

37. Simonis et al., "Structural Change." See also Archibugi and Nijkamp, "Ecological Modernization of Industrial Society."

38. Amory B. Lovins, "Energy, People and Industrialization." Interaction Council, High-Level Expert Group on Ecology and Energy Options, Montreal, April 29–30, 1989, Rocky Mountain Institute, January 1989.

39. "Cleaning Up After Communism." *The Economist* 314 February 17, 1990: 54–56. See also "Poison in the East," a five-part series on pollution in Eastern Europe, *Boston Globe*, December 17–21, 1989.

40. Jose Goldenberg, Thomas B. Johannson, Amalya K. N. Redday, and Robert H. Williams, *Energy for Development*, Washington, D.C.: World Resources Institute, 1985.

41. "Wasting Opportunities: President Bush Needs Only One Strategy for Energy: Save It." *The Economist* 317 (December 22, 1990): 14.

42. William D. Ruckelshaus, "Toward a Sustainable World." *Scientific American* 261 (September 1989): 166–74.

43. Jim MacNeill and Bob Munro, "Moving from the Margin to the Mainstream: 1972–1992." *EcoDécision* (Spring 1991).

Chapter 2

1. L. G. Uy, "Combatting the Notion of Environment as Additionality: A Study of the Integration of Environment and Development." Hobart, Tasmania: Center for Environmental Studies, University of Tasmania, 1985.

2. D. H. Meadows, D. L. Meadows, J. Randers, and W. Behrens, *The Limits to Growth*. London: Earth Isaland, 1972. See also H. Daly, "The Economic Growth Debate: What Some Economists Have Learned and Many Have Not." *Journal of Environmental Economics and Management* 14 (December 1987): 323–36.

3. United Nations, *Proceedings of the United Nations Conference on the Human Environment*. Stockholm, Sweden: United Nations, 1972.

4. Organization for Economic Cooperation and Development, *The State of Environment, 1985*. Paris: OECD, 1985.

5. "Cleaning Up After Communism." *The Economist* (February 17, 1989). See also "Poison in the East," *Boston Globe*, December 17–21, 1989.

6. The World Commission for Environment and Development, *Our Common Future*. Oxford: Oxford University Press, 1987.

7. Martin W. Holdgate, Mohammed Kassas, and Gilbert F. White, *The World Environment 1972–1982*. A report by the United Nations Environment Programme. Dublin: UNEP, Tycooly International Publishers Ltd., 1982.

8. See for example, *Economic Declaration*, Summit of the Arch, July 16, 1989. Quai d'Orsay, Paris; *Declaration of the Hague*, Environmental Summit, the Hague, March 11, 1989. Ministry of Housing, Physical Planning and the Environment, the Netherlands; *The Langkawi Declaration*, Commonwealth Heads of Government, Kuala Lumpur, October 18–24, 1989. Commonwealth Secretariat, London; *The Noordwijk Declaration on Climate Change*, November 6–7, 1989. Ministry of Housing, Physical Planning, and the Environment, the Netherlands; *Moscow Declaration of the Global Forum on Environment and Development for Human Survival*, January 1990. Ministry of Foreign Affairs, Moscow.

9. Mark Kosmo, *Money to Burn? The High Cost of Energy Subsidies*. Washington, D.C.: World Resources Institute, 1987.

10. Robert Repetto, *The Forests for the Trees? Government Policies and the Misuse of Forest Resources*. Washington, D.C.: World Resources Institute, May 1988.

11. Robert Repetto, *Paying the Price: Pesticide Subsidies in Developing Countries*, Washington, D.C.: World Resource Institute, 1985.

12. Robert Repetto, *Appropriate Incentives in Public Irrigation Systems*. Washington, D.C.: World Resource Institute, 1986.

13. Senate of Canada, *Soil and Risk: Canada's Eroding Future*. Ottawa: Government of Canada, 1984.

14. *Food 2000, Global Policies for Sustainable Agriculture*, the report of the Advisory Panel for Sustainable Food Security, Agriculture, Forestry and Environment to the World Commission on Environment and Development. London: Zed Books, 1987. See also *Our Common Future*, and World Bank, *World Development Report, 1986*, Washington, D.C.: World Bank.

15. "Panda Policy: Farmers Should Be Paid for Protecting the Environment." *The Economist* 317 (November 10, 1990): 17; and Cathal O'Connor, "Paying the Price for a Green and Pleasant Land." *The Independent*, October 28, 1990.

16. Robert Repetto, *Public Policies and the Misuse of Forest Resources*, Cambridge University Press, 1988. See also Repetto, *The Forests for the Trees?*

17. Jim MacNeill, André Saumier, John Cox, and David Runnalls, *CIDA and Sustainable Development*. Ottawa: Institute for Research on Public Policy, 1989.

18. Michel Potier, "Towards a Better Integration of Environmental, Economic and

Other Government Policies." Report on the 1989 Technology Transfer Conference, Toronto, November 20–21, 1989. Paris: OECD,

19. Mark Kosmo, *Money to Burn?*

20. Ian Brown, "Energy Subsidies in the United States," *Energy Pricing: Regulation, Subsidies and Distortion.* Surrey Energy Economics Centre Discussion Paper No. 38. Guildford, England: University of Surrey, March 1989.

21. Between 1984 and 1989, Canadian federal expenditures to promote energy efficiency fell from more than $400 million to less than $40 million.

22. R. C. Cavanagh, "Global Warming and Least-Cost Energy Planning." *Annual Review of Energy* 14 (1989): 353–73.

23. William U. Chandler and Andrew K. Nicholls, "Assessing Carbon Emissions Control Strategies: A Carbon Tax or a Gasoline Tax?" Draft MS of American Council for an Energy-Efficient Economy, September 15, 1989.

24. Ronald H. Coase, "The Problem of Social Cost." *Journal of Law and Economics* 3 (1960): 1–44; J. H. Dales, *Pollution, Property and Prices.* University of Toronto Press, 1968; William J. Baumol and Wallace E. Oates, *The Theory of Environmental Policy.* Englewood Cliffs, N.J.: Prentice Hall, 1975.

25. Garret Hardin, "The Tragedy of the Commons." In *Economics, Ecology and Ethics, Essays Toward a Steady-State Economy*, ed. Herman Daly. San Francisco: W. H. Freeman, 1980, pp. 100–20.

26. Joel S. Swisher and Gilbert M. Masters, "International Carbon Emission Offsets: A Tradeable Currency for Climate Protection Services," Technical Report 309, Dept. of Civil Engineering, Stanford University, February 28, 1989. President George Bush, Address to Intergovernmental Panel on Climate Change, Washington, D.C., February 1990.

27. Reference Project 88, *An Environmental Policy Agenda for the Next President.* Prepared under the auspices of Senators John Heinz and Tim Wirth, November 1988. Published by Project 88, Washington, D.C., 1988; and Robert N. Stavins, "Clean Profits: Using Economic Incentives to Protect the Environment." *Policy Review* 48 (Spring 1989): 58.

28. National Task Force on Environment and Economy. *The Green Report to the Canadian Council of Resource and Environment Ministers (CCREM).* CCREM, November 1987.

29. Government of Sweden, "Swedish Government Proposes Carbon-Dioxide Tax." Stockholm: Swedish Ministry of Environment and Energy, February 6, 1990. The legislation imposes a 23.46 percent value-added tax (VAT) on energy fuels, a levy of SKr 0.25 (4.1 cents) per kilo on carbon dioxide emissions (about 7 times the Finnish tax), and a levy of SKr 30 (about $4.90) per kilo of sulphur emissions. The VAT on motor fuels will also discriminate against lead and in favor of fuels with a lower content of sulpher and aromatic emissions.

30. Government of Sweden, *The Swedish Budget, 1990/91.* Summary published by Ministry of Finance.

31. The Finnish levy amounts to about $6.10 per tonne of carbon emissions. In all, green taxes and user charges will increase total tax revenue by some 1 percent. "Green Taxes, Where There's Muck There's Brass." *The Economist* 314 (March 17, 1990): 46–47.

32. Potier, "Towards a Better Integration of Environmental, Economic and Other

Government Policies."

33. Repetto, *The Forests for the Trees?*

34. Potier, "Towards a Better Integration of Environmental, Economic and Other Government Policies."

35. E. J. Mishan, *The Costs of Economic Growth.* London: Staples Press, 1967; D. W. Pearce, *The Economics of Natural Resource Depletion.* London: MacMillan, 1975; H. E. Daly, *Steady State Economics.* San Fransico: Freeman, 1977

36. *Our Common Future*; and *CIDA and Sustainable Development.*

37. The House of Commons of Canada, Bill C-78,"An Act to Establish a Federal Environmental Assessment Process." Ottawa: Canadian Government Publishing Centre, Supply and Services Canada, June 18, 1990.

38. *Our Common Future.*

39. Ibid. See also Jim MacNeill, *Testimony Before the Standing Committee on Environment, Minutes of Proceedings and Evidence of the Standing Committee on Environment.* Ottawa: House of Commons, May 25, 1989.

40. A Canadian Parliamentary Committee has recently recommended that an environmental "auditor general" be established by Parliament to undertake a continuing audit of all federal agencies and of the environmental implications of their policies and budgets. See Standing Committee on the Environment, *No Time To Lose: The Challenge of Global Warming.* Ottawa: Queen's Printer for Canada, October 1990.

41. The World Commission on Environment and Development, *Mandate for Change.* Geneva, August 1984.

Chapter 3

1. In the *Moscow Declaration of the Global Forum on Environment and Development for Human Survival*, January 1990, world leaders agreed to "return home and plant a tree." See also Office of the President, *Budget of the United States Government, Fiscal Year 1991.* IIIF, "Protecting the Environment."

2. A. H. Westing, ed., *Global Resources and International Conflict: Environmental Factors in Strategic Policy and Action.* New York: Oxford University Press, 1986.

3. Project Paper for Haiti Agroforestry Outreach Project (project no. 521–0122). Washington, D.C.: U.S. Agency for International Development, 1981.

4. U.S. Man and Biosphere Secretariat, "Draft Environmental Profile of El Salvador." Washington D.C.: U.S. Agency for International Development Bureau of Science and Technology, April 1982.

5. Relief and Rehabilitation Commission, "Drought and Rehabilitation in Wollo and Tigrai." Addis Ababa, 1975.

6. "Cleaning Up After Communism." *The Economist* 314 (February 17, 1990):54–56.

7. Malin Falkenmark, "New Ecological Approach to the Water Cycle: Ticket to the Future." *Ambio* 13. (1984): pp 152–160;

8. Joyce R. Starr and Daniel C. Stoll, *The Politics of Scarcity: Water in the Middle East.* Center for Strategic and International Studies. Boulder, Colo. and London: West-

view Press, 1988

9. Norman Myers, *Not Far Afield: U.S. Interests and the Global Environment.* Washington D.C.: World Resources Institute, June 1987.

10. Note the current crisis in the Canadian East Coast fisheries where stocks have collapsed because elected ministers have refused to establish the low quotas in Canadian waters recommended by scientists; instead, they point to foreign fleets in international waters as the real villians. See "Tough Fishing Limits Slip Through the Net." *New Scientist* 5 (January 1991): 14.

11. The July 1990, G7 summit in Houston provides a case in point. The seven northern leaders readily agreed to give priority to a convention on tropical deforestation, which is focused on the South but, mainly because of U.S. opposition, failed to agree on measures to reduce the use of fossil fuels, which would be focused on the North. See *Declaration of the Houston Summit*, July, 18, 1990; and "President Bush Lauds Results of Houston Summit." Transcript of President Bush's post-summit press conference. Washington, D.C.: White House, office of the press secretary, February 5, 1990.

12. Total Impact of a Community = Population $\times$ (Activity per Cap) $\times$ Impact per unit Activity. This useful accounting identity, presented by P. R. Ehrlich and J. P. Holdren in *Science* 171 (1971): 1212, holds that the instantaneous growth rate of environmental impact is the sum of the instantaneous growth rates of the three factors.

13. Although a measured activity (say, consumption of a hamburger) may occur in one place, its associated impact can be spread over many places: the rangeland where the beef was fed (maybe sustainable, maybe not; maybe on land deforested for grazing, maybe not), the rural landfill where the packaging waste will be discarded, the region where a dammed river provided the energy for cooking, the oilfield depleted for energy and petrochemical feedstocks to make the plastics used in packaging, and worldwide damage from ozone depletion caused by the CFC blowing agent used to manufacture the packaging.

14. See "The Problem of Interregional Trade" in William E. Rees, "A Role for Environmental Assessment in Achieving Sustainable Development." *Environmental Impact Assessment Review* 8 (1988): 271–72.

15. World Resources Institute. *World Resources 1988–89: An Assessment of the Resource Base that Supports the Global Economy.* New York: Basic Books, 1989.

16. U.N. General Assembly 44/228/ of December 22, 1989.

17. Omar Sattaur, "Convention Breaks Down Over Protecting Gene Pool." *New Scientist* (December 15, 1990): 00–00.

18. President George Bush, "Address to Intergovernmental Panel on Climate Change," Washington, D.C., February 1990.

19. UNEP/WMO, *Amendments to the Montreal Protocol on Substances that Deplete the Ozone Layer.* Adopted at London, June 29, 1990. Geneva: UNEP/WMO, 1990. The London Amendments added several CFCs to the list and calls for a complete phaseout by 2000. The amendments also call for a phaseout of halons by 2000; and added two key substances—methyl choloroform and carbon tetrachloride—to the list, calling for a 100 percent phaseout of the former by 2005 and the latter by 2000.

20. Valerie E. D. Heskins. "The Greening of the Summit: the Group of Seven Industrialized Democracies and the Environment Issue." Seven Power Summit Project, Center for Environmental Studies, University of Toronto, 1990.

21. John Kirton, "Sustainable Development at the Houston Seven Power Summit." Center for International Studies, University of Toronto.

22. See Alain Giguere, testimony before Parliamentary Forum on Global Climate Change, April 23, 1990. Ottawa: Queen's Printer for Canada, 1990.

23. "People Around the World Worry Deeply About Pollution, Survey for UN Finds." (Montreal) *Gazette*, May 10, 1989. Reports on a Louis Harris survey for UNEP of 14 countries, including 11 in the Third World.

24. *The Langkawi Declaration*, Commonwealth Heads of Government, October 18–24, 1989.

25. Robert Mugabe, Prime Minister of Zimbabwe, address to U.N. General Assembly October, 19, 1987.

26. Jozef Goldblat, ed., *Non-Proliferation: The Why and the Wherefore*. London: Taylor and Francis, 1985.

27. R. D. Lipschutz, "The Political Economy of Strategic Materials." Berkeley, Calif.: Pacific Institute for Studies in Development, Environment and Security, 1988.

28. R. D. Lipschutz and J. P. Holdren, "Crossing Borders: Resource Flows, The Global Environment, and International Security." Berkeley, Calif.: Pacific Institute for Studies in Development, Environment and Security, 1989.

29. See World Commission on Environment and Development, *Our Common Future*, chapter 11, Oxford: Oxford University Press, 1987.

30. Jessica Tuchman Mathews, "Redefining Security." *Foreign Affairs*, Spring 1989, Vol 68 No 2, pp 162–177

31. Organization for Economic Cooperation and Development, *Economic and Ecological Interdependence*. Paris: OECD, 1981.

Chapter 4

1. D. Lashof and D. Tirpak, *Policy Options for Stabilizing Global Climate*. EPA Draft Report to the U.S. Congress, February 1989.

2. Intergovernmental Panel on Climate Change, *Scientific Assessment of Climate Change*. Report to IPCC from Working Group 1, third draft. Bracknell, U.K.: Meteorological Office, May 2, 1990.

3. Second World Climate Conference, "Final Conference Statement, Scientific/Technical Sessions, 7 November, 1990," par. 8. Geneva: World Meteorological Organization and United Nations Environment Program, 1990.

4. IPCC, "Potential Impacts of Climate Change." Report prepared for IPCC by Working Group II, June 1990. Geneva: World Meteorological Organization and United Nations Environment Program, 1990.

5. Second World Climate Conference, "Ministerial Declaration of the Second World Climate Conference, 7 November, 1990," par. 8. Geneva: World Meteorological Organization and United Nations Environment Program, 1990.

6. Second World Climate Conference, "Formulation of Response Strategies. Report Prepared for IPCC by Working Group III, June 1990. Geneva: World Meteorological Organization and United Nations Environment Program, 1990.

7. "Ministerial Declaration of the Second World Climate Conference," par. 12.

8. Government of Japan, "Action Program to Arrest Global Warming: Decision made by the Council of Ministers for Global Environment Conservation." Tokyo: Government of Japan, October 23, 1990.

9. The ministers recognized these "differences in approach and in starting point in the formulation of the targets." "Ministerial Declaration of the Second World Climate Conference," par. 12.

10. See statements by Brazil, China, and India to Ministerial Conference. *Proceedings of the Second World Climate Conference.* Geneva: World Meteorological Organization and United Nations Environment Program, 1990.

11. Kirk Smith, "Global Warming and the National Debt." Hawaii: Environment and Policy Institute, East–West Center, 1989.

12. Conference Statement; op cit.

13. The World Commission on Environment and Development, *Our Common Future.* Oxford, Oxford University Press, 1987.

14. UNEP, "General Assessment of Progress in the Implementation of the Plan of Action to Combat Desertification 1978–1984." Nairobi: UNEP, 1984.

15. James Gustav Speth, "Coming to Terms: Toward a North–South Bargain for the Environment." Washington, D.C.: World Resources Institute.

16. The term "global bargain" focuses on international bargains among often far-flung partners, generally crossing the North–South and East–West divides.

17. The World Bank, *World Development Report. 1989.* Washington, D.C.: The World Bank.

18. World Resources Institute, *Natural Endowments, Financing Resource Conservation for Development.* International Conservation Financing Project, commissioned by the United Nations Development Program, World Resources Institute, Washington, D.C.

19. Apart from the World Commission on Environment and Development (*Our Common Future.* Oxford: Oxford University Press, 1987), these include World Resources Institute, Washington, D.C. (e.g., *Natural Endowments, Financing Resource Conservation for Development*), the Institut fur Europaische Umweltpolitik, Bonn, Germany (e.g., Ernst U. von Weiszacker, "Regulatory Reform and the Environment: The Case for Environmental Taxes"), the Institute for Research on Public Policy, Ottawa, Canada (e.g, Jim MacNeill, André Saumier, David Runnalls, John Cox, *CIDA and Sustainable Development.* Ottawa: Institute for Research on Public Policy, 1989), the Netherlands Government (e.g., McKinsey & Company, "Protecting the Global Atmosphere: Funding Mechanisms." Report to the Ministerial Conference on Atmospheric Pollution and Climate Change, Noordwijk, the Netherlands, November 1989), and the Royal Institute of International Affairs (e.g., Michael Grubb, "The Greenhouse Effect: Negotiating Targets." London: Royal Institute of International Affairs, November 1989.

20. David Runnalls, "The Grand Bargain." Remarks to the Canadian Institute for International Affairs, Ottawa, Canada: Institute for Research on Public Policy, May 3, 1989. James Gustav Speth, "Bargaining for the Future, Debt Relief and Climate Protection." Washington, D.C.: World Resources Institute, 1990.

21. UNEP/WMO, *Amendments to the Montreal Protocol on Substances that Deplete the Ozone Layer.* Adopted at London, June 29, 1990. Geneva: UNEP/WMO, 1990.

22. McKinsey & Company, "Protecting the Global Atmosphere."

23. Dr. Amil Markandya, "The Costs to Developing Countries of Entering the Montreal Protocol: A Synthesis Report." UNEP, May 1990.

24. The studies also cover halons, carbon tetrachlorides, and methyl chloroform.

25. This work was initiated at a UNEP workshop in Nairobi, August, 21–25, 1989. The project was further advanced at an EPA-sponsored workshop in Washington, D.C., January 15–17, 1990.

26. *Our Common Future*, pp. 314, 319.

27. MacNeill et al., *CIDA and Sustainable Development*, chapter 2.

28. Report of the Secretary General, "Technical and Economic Aspects of International River Basin Development." UN E/C.7/35, New York, 1972. See also, "Experiences in the Development and Management of International River and Lake Basins." Proceedings of the UN Interregional Meeting of the International River Organizations held at Dakar, Senegal, May 1981. New York: United Nations, 1983.

29. This idea comes from Robert Williams, Center for Energy and Environmental Studies, Princeton University.

30. See *Our Common Future*; and; Jose Goldenberg, Thomas B. Johannson, Amulya K. N. Reddy, and Robert H. Williams, *Energy for Development*. Washington, D.C.: World Resource Institute, 1985.

31. The DPA Group, Inc., *Study on the Reduction of Energy-Related Greenhouse Gas Emissions*. Commissioned by the Ontario Ministry of Energy (with support from all the federal and provincial energy departments of Canada). Prepared by the DPA Group, Inc., in association with CH₄ International Ltd., RCG/Hagler, Bailly, Inc., Steven G. Diener and Associates Ltd., Ontario Ministry of Energy, March 1989.

32. *Environmental Action Plan of Netherlands*. Electricity and Gas Distribution Companies (VEEN, VEGIN, VESTIN), April 1990.

33. K. Blok, E. Worrell, R. A. W. Albers and R. F. A. Cuelenaere, "Data on Energy Conservation Techniques for the Netherlands." University of Utrecht, 1990; Tim Jackson and Simon Roberts, "Getting Out of the Greenhouse, An Agenda for UK Action on Energy Policy." London: Friends of the Earth, December 1989.

34. Tim Jackson, "The Least Cost Greenhouse Planning: Supply Curves for Global Warming Abatement." *Energy Policy* 19 (January/February 1991).

35. Robert H. Williams, "Low-Cost Strategies for Coping with CO_2 Emission Limits" (a critique of Alan S. Manne and Richard G. Richels, "CO_2 Emission Limits: An Economic Cost Analysis for the U.S.A." *The Energy Journal*, November 1989. Center for Energy and Environmental Studies, Princeton University, December 1989. And B. Bodlund, (Vattenfall), E. Mills (U. of Lund), T. Karlsson (Vattenfall), and T. Johansson (U. of Lund), *The Challenge of Choices: Technology Options for the Swedish Electricity Sector in Electricity: Efficient End-Use and New Generation Technologies and the Planning Implications*, T. Johannson, B. Bodlund and R. Williams, eds.. Sweden: Lund University Press, 1989.

36. Brita Bye, Torstein Bye, and Lorents Lorentsen, "SIMEN: Studies of Industry, Environment and Energy Towards 2000." Central Bureau of Statistics, Discussion Paper No. 44, Oslo, August 1989.

37. Netherlands Central Planning Bureau, in *National Environmental Policy Plan of the Netherlands*, Ministry of Housing, Physical Planning, and the Environment, the Netherlands, 1989. Directorate General for Energy, Commission of the European Communities, *Energy for a New Century: The European Perspective—Major Themes in Energy Revisited*. Brussels, May 1990.

38. McKinsey & Company, "Protecting the Global Atmosphere."

39. Ibid.

40. Note that this concerns international funding requirements, not the total cost to industrialized and developing societies.

41. Many OECD countries saw expenditures on environmental pollution alone (i.e., not including expenditures on resource management) rise from about 1.3 percent of GNP in 1970 to somewhere between 1.5 percent and 2 percent around the end of the decade. *Our Common Future*, p. 335.

42. See, for instance, International Energy Agency, "Energy and the Environment: Policy Overview." Paris: IEA/OECD, 1989.

43. "Bergen Ministerial Declaration on Sustainable Development in the ECE Region." Bergen, Norway: ECE, May 16, 1990.

44. Intergovernmental Panel on Climate Change, *Scientific Assessment of Climate Change*. Report to IPCC from Working Group 1, third draft. Bracknell, U.K.: Meteorological Office, May 2, 1990.

45. John Valorzi, "U.S. Rejects Call for Green Fund." *Canadian Press*, July 12, 1990. See also *Declaration of the Houston Summit*, July 18, 1990.

46. Manne and Richels, "CO_2 Emission Limits."

47. Robert H. Williams, "Low-Cost Strategies for Coping with CO_2 Emission Limits." Center for Energy and Environmental Studies, Princeton University, December 1989.

48. William D. Nordhaus, "To Slow or Not to Slow: The Economics of the Greenhouse Effect," February 5, 1990; and "The Economics of the Greenhouse Effect," a paper presented to the 1989 meetings of the International Energy Workshop and the MIT Symposium on Environment and Energy.

49. Peter Passell, "Curing the Greenhouse Effect Could Run into the Trillions," *New York Times*, November 19, 1989, p. 1.

50. The Second World Climate Conference recommended that "consideration should be given to the need for funding facilities, including the proposed World Bank Facility, a clearing house mechanism and a new possible international fund. Such funding should be related to the implementation of the framework convention on climate change and any other related instruments that might be agreed upon." "Ministerial Declaration of the Second World Climate Conference," par. 17.

51. As of January 1991, China and India had not done so.

52. The Second World Climate Conference stated that "To enable developing countries to meet incremental costs required to take the necessary measures to address climate change and sea-level rise, consistent with their development needs, we recommend that adequate and additional financial resources should be mobilized and best available environmentally sound technologies transferred expeditiously on a fair and most favorable basis." "Ministerial Declaration of the Second World Climate Conference," par. 15.

53. For example, the Hague, March 1989, and Helsinki, May 1989. The 1989 Noordwijk Conference essentially confirmed this suggestion, but chose to speak of funding mechanisms to broaden the range of practical options.

54. John Valorzi, "U.S. Rejects Call for Green Fund." *Canadian Press*, July 12, 1990. Also *Declaration of the Houston Summit*, July 1990.

55. World Resources Institute, *Natural Endowments: Financing Resource Conservation for Development*.

56. Ibid.

57. Anthony Cassils and Geza P. dePatrallyay, "Let's Give Businesses an Incentive

They Understand." *Globe and Mail*, October 21, 1988.

   58. World Resources Institute, *Natural Endowments*.

   59. A. R. Dobell and E. A. Parson, "A World Atmosphere Fund," *Policy Options*, November 1988, proposes an international emission permit scheme in which some permits are given to low-income countries on a per capita basis, and others are auctioned to raise funds for international atmosphere protection. M. J. Grubb, Royal Institute for International Affairs, proposes free distribution based on adult population. The American delegation to the IPCC has also been informally advancing an approach to greenhouse reduction based on distribution of tradable permits. Discussion papers for February 3, 1990, seminar for IPCC participants, U.S. Department of State.

   60. Michael Grubb, "The Greenhouse Effect: Negotiating Targets." London: Energy and Environment Programme, the Royal Institute of International Affairs, 1989.

   61. This commitment was part of the U.N.'s International Development Strategy for the 2nd Development Decade. See General Assembly Resolution 2626 (XXV), October 24, 1970.

   62. The World Bank, *World Development Report, 1989*.

   63. McKinsey & Company, *Protecting the Global Atmosphere*.

   64. "Ministerial Declaration of the Second World Climate Conference."

   65. "President Bush Lauds Results of Houston Summit." Transcript of President Bush's post-summit press conference. Washington, D.C.: White House, office of the press secretary, February 5, 1990.

   66. "Green Taxes, Where There's Muck There's Brass," *The Economist*, 314 (March 17, 1990): 46–47.

Chapter 5

   1. See *Environmental Protection and Sustainable Development: Legal Principles and Recommendations adopted by the WCED Experts Group on Environmental Law*. London: Graham & Trotman/Martinus Nijhoff, 1987. For a summary of the legal principles, see Annex 1 of *Our Common Future*. Oxford: Oxford University Press, 1987.

   2. *Our Common Future*, opening sentence.

   3. *Report of the International Conference on The Changing Atmosphere: Implications for Global Security*, Toronto, Canada, June 27–30, 1988. Toronto: Environment Canada, 1988.

   4. For another discussion of the elements of a convention, see William A. Nitze, *The Greenhouse Effect: Formulating a Convention, Energy and Environment Programme*. London: Royal Institute of International Affairs, 1990.

   5. Richard E. Benedick, "The Ozone Protocol: A New Global Diplomacy." *Conservation Fund Letter*, no. 4, 1989.

   6. Mostafa Tolba argues for parallel negotiating tracks but falls short of insisting that completion of the framework convention should await the successful negotiation of the protocols. See "A Step by Step Approach to Protection of the Atmosphere." Address

to Law and Policy Experts, Ottawa, February 20–22, 1989. published in *International Environmental Affairs* 1:4 (Fall 1989):304–08.

7. R. Bertrand, *Some Reflections on Reform of the United Nations.* Geneva: U.N. Joint Inspection Unit, 1985; see also *Our Common Future,* chapter 12.

8. Others could also be mentioned: e.g., WHO, which has enjoyed repeated successes in environmental health and disease control; UNESCO, which has maintained quality programs in oceans and water management, led the establishment of biological reserves and World Heritage sites, and led and supported the Man and Biosphere (MAB) program through a number of extremely difficult years.

9. In 1987, the U.N.'s Administrative Coordinating Committee reported that the annual investment required to meet the costs of implementing the PACD was about one-tenth of the annual losses incurred by desertification. Nonetheless, it found that the mechanism set up to secure additional funding for the program to be "marginal and inadequate."

10. Ibid.

11. *The Declaration of the Hague,* Environmental Summit, the Hague, March 11, 1989.

12. Bertrand, *Some Reflections on Reform of the United Nations.*

13. McKinsey & Company, "Protecting the Global Atmosphere: Funding Mechanisms." Report to the Ministerial Conference on Atmospheric Pollution and Climate Change, Noordwijk, the Netherlands, November 1989.

14. Jim MacNeill and Bob Munro, "From the Margins to the Mainstream: 1972–1992." *EcoDécision,* Spring 1991.

About the Authors

Jim MacNeill is secretary general of the World Commission on Environment and Development (Brundtland Commission) and the principal architect and main author of its acclaimed report, *Our Common Future*. He is president of MacNeill & Associates and a senior fellow at the Institute for Research on Public Policy in Ottawa. Prior to joining the World Commission in 1984, he was for six years director of environment for the OECD. Mr. MacNeill served for two years as Canada's commissioner general and ambassador extraordinary and plenipotentiary for the 1976 United Nations Conference of Human Settlements (Habitat) held in Vancouver. Between 1972 and 1976 he was assistant permanent secretary and then permanent secretary (deputy minister) in the Canadian Ministry of State for Urban Affairs. Earlier, he was a special advisor on the constitution and environment in Prime Minister Pierre Trudeau's office. While there, he authored an original work published under the title "Environmental Management in Canada" and played a leading role in Canada's preparations for the 1972 U.N. Conference on the Human Environment in Stockholm, Sweden. Still earlier, he managed federal responsibilitiees for Canada's water and renewable resources, including Canada's marine resources. Mr. MacNeill began his career in Saskatchewan in 1952 as an economist with the province's Economic Advisory and Planning Board. He later became executive director of the South Saskatchewan River Development Commission, and served as the first executive director of the Saskatchewan Water Resources Commission. Mr. MacNeill has degrees in science (mathematics and physics) (1949), and mechanical engineering (1958) and an honorary

LL.D. from the University of Saskatchewan; and a graduate diploma in economics and political science from the University of Stockholm (1951). He is the author of numerous articles, books, and reports, and is a member of the board of several institutes in Canada and the United States.

Pieter Winsemius is a director of McKinsey & Company in Amsterdam, where he also worked prior to serving as Dutch minister of housing, physical planning, and environment from 1982 to 1986. As minister he chaired the OECD Ministerial Conference on Environment and Economics held in Paris in 1984. Mr. Winsemius was educated in physics at Leiden University in the Netherlands. While working on his doctoral dissertation (which he completed in 1973), he was employed by the Foundation for Fundamental Research on Matter (F.O.M.). After obtaining an M.B.A. from Stanford University, Mr. Winsemius joined McKinsey & Company, where he was elected a partner in 1980. His current responsibilities at McKinsey include a special focus on the strategic and organizational implications of technological developments and environmental issues, specifically, an international proposal to clean up the Rhine River, chemical waste management and, most recently, global warming. Mr. Winsemius is the author of numerous publications on the environment, including *Guest in Our Own Home* (1986, in Dutch), which was widely acclaimed in the environmental and business press. He currently serves as chairman of the Netherlands' Society for Nature Conservancy.

Taizo Yakushiji is professor of technology and international relations at the Graduate School of Policy Science at Saitama University, Japan. He was educated at Keio University (B.S. in electrical engineering), the University of Tokyo (B.A. in history and philosophy of science), and the Massachusetts Institute of Technology (Ph.D. in political science). He studied in the United States as a Fulbright scholar and Ford Foundation fellow (1970–75). He was a visiting senior research associate at the Berkeley Roundtable on the International Economy and the Department of Political Science of the University of California at Berkeley from 1984 to 1985. He was selected as one of the 1988 Young Leaders of Asia by the U.S.–Asia Institute in Washington, D.C. Mr. Yakushiji has participated in many international research projects and conferences on the automobile industry, science and technology policies, international cooperation, dual-use technologies, and telecommunication policies. He has written numerous articles and books in both English and Japanese. His publications include *Reshuffling Firms for Technology?* (1984), *The American and Japanese Auto Industries in Transition* (1984), *Policy, Corporate Ideology and the Auto Industry* (1986), *The Politicians and Bureaucrats* (1987, Japanese), *The Politics of High Technology* (1988), *The Techno-Hegemony* (1989, Japanese), and *Europe and Japan Facing High Technologies* (1989).

Index